Augsburger Schriften zur Mathematik, Physik und Informatik
Band 19

herausgegeben von:
Professor Dr. F. Pukelsheim
Professor Dr. W. Reif
Professor Dr. D. Vollhardt

Bibliografische Information der Deutschen Nationalbibliothek

Die Deutsche Nationalbibliothek verzeichnet diese Publikation in der
Deutschen Nationalbibliografie; detaillierte bibliografische Daten sind
im Internet über http://dnb.d-nb.de abrufbar.

ISBN 978-3-8325-3088-4
ISSN 1611-4256

Logos Verlag Berlin GmbH
Comeniushof, Gubener Str. 47,
10243 Berlin
Tel.: +49 030 42 85 10 90
Fax: +49 030 42 85 10 92
INTERNET: http://www.logos-verlag.de

A posteriori error estimation for hybridized mixed and discontinuous Galerkin methods

Dissertation zur Erlangung des Doktorgrades der
Mathematisch-Naturwissenschaftlichen Fakultät der
Universität Augsburg

vorgelegt von Johannes L. Neher

November 2011

Erstgutachter: Prof. Dr. R. H. W. Hoppe, Universität Augsburg

Zweitgutachter: Prof. Dr. M. A. Peter, Universität Augsburg

Drittgutachter: Prof. Dr. G. Kanschat, Texas A & M University

Mündliche Prüfung: 21. Dezember 2011

Contents

Chapter 1

Introduction

In this thesis, we study adaptive hybrid mixed and discontinuous Galerkin methods, as well as H^1-conforming and nonconforming finite element methods for the approximation of linear second order partial differential equations (PDE) based on a simplicial triangulation of the underlying computational domain. In order to get a direct representation of the flux of the solution, the second order PDE can be reformulated as a system of first order PDEs.

The classical mixed finite element methods, (initiated in [49]), use finite dimensional subspaces of the function spaces involved in the dual mixed formulation to approximate these quantities. See also [12, 13] and the references therein. These methods have been studied intensively and extended in order to improve the accuracy of the methods in suitable norms (see for instance [4, 12, 11, 41]).

The requirement of conforming approximation spaces can be relaxed in favor of additional penalty terms in the variational formulation, which penalize present discontinuities in the primal and the dual variable. Discontinuous Galerkin (DG) methods are a realization of this idea. Since the involved functions do not have to fulfill the continuity constraints of conforming methods, DG methods inherit algorithmic flexibility – in particular for the realization of higher order polynomial ansatz spaces. Therefore, they have become a popular alternative technique to numerically approximate the solution of the first order PDE system. We refer to [22] for an introduction into DG methods for different kinds of applications. For elliptic equations the most widely used schemes are the Interior Penalty DG (IPDG) and the Local DG (LDG) methods. An a priori analysis of these DG methods has been provided, e.g. in [5, 21, 6, 8].

In contrast to the numerical flexibility, DG methods are disadvantageous, because a relatively large linear algebraic system has to be solved in comparison to standard H^1-conforming finite element methods. Hybridization is an effective technique to significantly decrease the size of the linear system. After having been initially proposed in [28] in the context of equilibrium analysis in structural mechanics, it has been adapted for the mixed formulation of the diffusion equation, developed continuously, and used to make clear the connections between different methods in [4, 12, 23, 24]. In recent years the idea of hybridization has been extended to various other kinds of finite element methods, including DG methods, in [25, 26].

Adaptive finite element methods on the basis of reliable a posteriori error estimators are well understood for H^1-conforming methods, see e.g. [55, 2, 44, 43, 20] and the references therein.

Adaptive mixed, mixed hybrid, and DG methods using residual based a posteriori error estimators have been studied for example in [3, 10, 14, 35, 60, 18, 45, 39]. Other error concepts like local averaging, hierarchical estimates, or functional type error estimates have been studied in [3, 60, 16, 37, 50, 57, 58]. Comparative studies of different adaptive methods can be found in [60, 1, 39]. A convergence analysis of certain adaptive DG methods has been provided in [38, 34, 9]. However, the author is not aware of any investigations of adaptive hybrid DG methods.

The thesis is organized as follows: In Chapter 2, we state the boundary value problem under consideration, introduce three different variational formulations thereof, discuss the question of well-posedness, and the relation between the three formulations. For the formulation which is used as a basis for discretization, we prove an explicit stability result. This is useful to obtain computable error estimates in Chapter 4. At the end of this chapter we introduce some notation related to the finite dimensional approximation of the boundary value problem using the concept of hybridization. Chapter 3 is devoted to the well-posedness of a general hybrid method and the application to mixed methods, DG-methods, H^1-conforming, and nonconforming methods. We recall earlier established results, concerning a priori estimates, and point out connections between the original methods and their hybridized versions. In the following chapter, we prove the reliability of an a posteriori error estimator suitable for all kinds of hybrid methods discussed in Chapter 3 and for any polynomial degree of the finite dimensional function spaces. The proof is based on the evaluation of the residual of the primal mixed formulation of the underlying partial differential equation, with respect to the computed approximation of the dual variable and a conforming approximation of the primal variable. Furthermore, efficiency is proved for some of the methods discussed in Chapter 3 up to constants, that may depend on the polynomial degree of the discrete spaces. All results will be given up to constants which do not depend on the local mesh size and the polynomial degree of the local finite dimensional spaces. In Chapter 5, a detailed documentation of numerical experiments will be given, after presenting an overview of the adaptive algorithm used to obtain the computational results. For a problem with smooth solution the a priori error estimates of Chapter 2 will be confirmed. Furthermore, all methods will be applied to test problems without smooth solutions, which highlight the advantages of adaptive hybrid methods in comparison to their non-adaptive counterparts. The theoretical results of Chapter 4 will be confirmed by the numerical results.

Chapter 2

Hybridized problem

2.1 Primal and dual mixed formulation.

We use standard notation from Lebesgue and Sobolev space theory [54].

In particular, given an open bounded polygonal domain $\Omega \neq \emptyset$ of two-dimensional Euclidian space $\mathbb{R}^2$ with boundary $\Gamma := \partial\Omega$, for $D \subseteq \Omega$ we refer to $L^p(D), 1 \leq p \leq \infty$ as the Banach spaces of p-th power integrable functions ($p < \infty$) and essentially bounded functions ($p = \infty$) on D with norm $\|\cdot\|_{0,p,D}$. $L^p_+(D)$ stands for the positive cone on $L^p(D)$. In case $p = 2$, the space $L^2(D)$ is a Hilbert space. Dropping the index $p = 2$, inner product and norm will be referred to as $(\cdot, \cdot)_{0,D}$ and $\|\cdot\|_{0,D}$, respectively. For $m \in \mathbb{N}_0$, we denote by $W^{m,p}(D)$ the Sobolev space with the norm

$$\|v\|_{m,p,D} := \begin{cases} (\sum\limits_{|\alpha| \leq m} \|D^\alpha v\|^p_{0,p,D})^{1/p} & , \text{if } p < \infty \\ \max\limits_{|\alpha| \leq m} \|D^\alpha v\|_{0,\infty,D} & , \text{if } p = \infty \end{cases},$$

where $\alpha = (\alpha_1, \alpha_2) \in \mathbb{N}_0^d$ is a multi-index with $|\alpha| := \alpha_1 + \alpha_2$. Further we refer to $|\cdot|_{m,p,D}$ as the associated semi-norms of $W^{m,p}(D)$. For $p < \infty$ and $s \in \mathbb{R}_+$, $s = m + \sigma$, $m \in \mathbb{N}_0$, $0 < \sigma < 1$, we denote by $W^{s,p}(D)$ the Sobolev space with norm

$$\|v\|_{s,p,D} := \left(\|v\|^p_{m,p,D} + \sum_{|\alpha|=m} \int_D \int_D \frac{|D^\alpha v(\mathbf{x}) - D^\alpha v(\mathbf{y})|^p}{|\mathbf{x} - \mathbf{y}|^{d+\sigma p}} \, d\mathbf{x} \, d\mathbf{y}\right)^{1/p}.$$

We refer to $W^{s,p}_0(D)$ as the closure of $C^\infty_0(D)$ in $W^{s,p}(D)$. For $s < 0$, we denote by $W^{-s,p}(D)$ the dual space of $W^{s,q}_0(D)$, where $1/p + 1/q = 1$. In case $p = 2$, the spaces $W^{s,2}(D)$ are Hilbert spaces. We will write $H^s(D)$ instead of $W^{s,2}(D)$ and refer to $(\cdot, \cdot)_{s,D}$ and $\|\cdot\|_{s,D}$ as the inner products and associated norms. For $v \in H^1(D)$, we denote by

$\gamma_0(v) \in H^{1/2}(\partial D)$ the trace of v on ∂D and note that $\gamma_0 : H^1(D) \rightarrow H^{1/2}(\partial D)$ is a surjective and continuous linear operator. For $u \in H^{1/2}(\partial D)$ and $v \in H^{-1/2}(\partial D)$, the pairing $\langle v, u \rangle_{\partial D}$ stands for the dual pairing between $H^{-1/2}(\partial D)$ and $H^{1/2}(\partial D)$. We further refer to $H(div; D) := \{\mathbf{q} \in L^2(\Omega)^2| \nabla \cdot \mathbf{q} \in L^2(D)\}$ as the Hilbert space with the graph norm

$$\|\mathbf{q}\|_{div;\Omega} := (\|\mathbf{q}\|_{0,\Omega}^2 + \|\nabla \cdot \mathbf{q}\|_{0,\Omega}^2)^{1/2}$$

and denote by $\nu_\Gamma(\mathbf{q}) := \mathbf{q} \cdot \mathbf{n} \in H^{-1/2}(\Gamma)$ the normal trace of $\mathbf{q} \in H(div; \Omega)$ on Γ, where $\mathbf{n}$ stands for the exterior unit normal vector on Γ. We note that the normal trace mapping $\nu_\Gamma : H(div; \Omega) \rightarrow H^{-1/2}(\Gamma)$ is a surjective linear continuous mapping.

We consider the reaction diffusion equation on a polygonal domain $\Omega \subset \mathbb{R}^2$ and Dirichlet boundary conditions

$$-\nabla \cdot a\nabla u + du = f \qquad\qquad \text{in } \Omega , \qquad\qquad (2.1a)$$

$$u = g_D \qquad\qquad \text{on } \partial\Omega , \qquad\qquad (2.1b)$$

where $f \in L^2(\Omega)$, $g_D \in H^{1/2}(\Gamma)$, $a = (a_{ij})_{i,j=1}^2$ is a symmetric, uniformly positive definite matrix-valued function with entries $a_{ij} \in L^\infty(\Omega)$, $1 \leq i, j \leq 2$, and $d \in L_+^\infty(\Omega)$.

2.1.1   Primal variational formulation

To solve this equation we turn to the variational formulation of the problem. After multiplying this equation with any function $w \in H_0^1(\Omega)$, integrating over the domain Ω and using integration by parts, we arrive at the primal variational formulation of the partial differential equation that reads: Find a function $u \in H^1(\Omega)$ which satisfies the boundary condition such that

$$(a\nabla u, \nabla w) + (du, w) = (f, w) \qquad\qquad \forall w \in H_0^1(\Omega) . \qquad\qquad (2.2)$$

To guarantee the unique solvability of (2.2) we identify the left hand side of (2.2) as a continuous and coercive bilinear form acting on u and w and subtract an extension $u_g \in H^1(\Omega)$ of g_D from the solution u to arrive at an equation with homogeneous Dirichlet boundary condition and apply the the Lax Milgram Lemma.

2.1.2   Primal mixed formulation

An alternative approach to finding solutions of (2.1) is to rewrite the second order equation as a system of first order partial differential equations by defining $\mathbf{q} := -a\nabla u$ as an extra variable.

$$\mathbf{q} + a\nabla u = 0 \ , \tag{2.3a}$$

$$\nabla\cdot\mathbf{q} + du = f \ . \tag{2.3b}$$

Based on (2.3), the primal mixed formulation of (2.1) amounts to the computation of $(u, \mathbf{q}) \in W \times \mathbf{V}$, where with $W_0 := H_0^1(\Omega)$ we have $W := \{g_D\} + W_0$ and $\mathbf{V} := L^2(\Omega)^2$, such that there holds

$$a_P(\mathbf{q}, \mathbf{v}) + b_P(u, \mathbf{v}) = l_1(\mathbf{v}) \qquad\qquad \forall \mathbf{v} \in \mathbf{V} \ , \tag{2.4a}$$

$$b_P(w, \mathbf{q}) - d_P(u, w) = l_2(w) \qquad\qquad \forall w \in W_0 \ . \tag{2.4b}$$

Here, the bilinear forms $a_P(\cdot, \cdot)$, $b_P(\cdot, \cdot)$ and $d_P(\cdot, \cdot)$ are defined according to

$$a_P(\mathbf{q}, \mathbf{v}) := (a^{-1}\mathbf{q}, \mathbf{v})_{0,\Omega}, \qquad b_P(u, \mathbf{v}) := (\nabla u, \mathbf{v})_{0,\Omega}, \qquad d_P(u, w) := (du, w)_{0,\Omega},$$

whereas the functionals $l_1 \in \mathbf{V}^*$ and $l_2 \in W_0^*$ are defined by

$$l_1(\mathbf{v}) := 0 \ , \qquad\qquad l_2(w) := (-f, w)_{0,\Omega} \ .$$

Denoting by $A_P : \mathbf{V} \to \mathbf{V}^*$, $B_P : \mathbf{V} \to W_0^*$ and $D_P : W \to W_0^*$ the operators associated with the bilinear form by means of

$$\langle A_P\mathbf{q}, \mathbf{v}\rangle_{\mathbf{V}^*,\mathbf{V}} := a_P(\mathbf{q}, \mathbf{v}) \qquad\qquad \mathbf{q}, \mathbf{v} \in \mathbf{V} \ ,$$

$$\langle B_P\mathbf{q}, w\rangle_{W_0^*,W_0} := b_P(w, \mathbf{q}) \qquad\qquad \mathbf{q}, \in \mathbf{V} \ , w \in W_0 \ ,$$

$$\langle D_Pu, w\rangle_{W_0^*,W_0} := d_P(u, w) \qquad\qquad u, w \in W_0 \ ,$$

we introduce an operator $\mathcal{L}_P : \mathbf{V} \times W \to \mathbf{V}^* \times W_0^*$ according to

$$\mathcal{L}_P(\mathbf{q}, u) := \begin{bmatrix} A_P & B_P^* \\ B_P & -D_P \end{bmatrix} \begin{bmatrix} \mathbf{q} \\ u \end{bmatrix} \ , (\mathbf{q}, u) \in \mathbf{V} \times W \ . \tag{2.5}$$

Then, the operator-theoretic formulation of (2.4) reads:

$$\mathcal{L}(\mathbf{q}, u) = (l_1, l_2)^T \ . \tag{2.6}$$

Theorem 2.1.1

The operator $\mathcal{L}_P : \mathbf{V} \times W_0 \to \mathbf{V}^* \times W_0^*$, as given by (2.5), is a bijective, continuous linear operator. Consequently for any given $l_1 \in \mathbf{V}^*$, $l_2 \in W_0^*$ the operator equation (2.6) admits a unique solution $(\mathbf{q}, u) \in \mathbf{V} \times W_0$. The solution depends continuously on the data, i.e., there exists a constant $C_P \geq 0$ such that

$$\|(\mathbf{q}, u)\|_{\mathbf{V} \times W_0} \leq C_P \|(l_1, l_2)\|_{\mathbf{V}^* \times W_0^*} \ . \tag{2.7}$$

We present a direct proof in analogy to the proof of Theorem 2.1 of [15].

Proof. Let $\|\cdot\|_a$ denote the problem dependent norm of the product space $X := \mathbf{V} \times W_0$, defined by

$$\|(\mathbf{q}, u)\|_a := (\|\mathbf{q}\|_{\mathbf{V},a}^2 + \|u\|_{W_0,a}^2)^{1/2} \ ,$$

where $\|\mathbf{q}\|_{\mathbf{V},a} := (a^{-1}\mathbf{q}, \mathbf{q})_{0,\Omega}^{1/2}$ and $\|u\|_{W_0,a} := ((a\nabla u, \nabla u) + (du, u)_{0,\Omega})^{1/2}$. For $(\mathbf{v}, w) := (\mathbf{q} + a\nabla u, -2u)$, we estimate

$$\begin{aligned}
\|(\mathbf{v}, w)\|_a^2 &= \|(\mathbf{q} + a\nabla u, -2u)\|_a^2 \\
&\leq (\|\mathbf{q}\|_{\mathbf{V},a} + \|a\nabla u\|_{\mathbf{V},a})^2 + 2(du, u) + 2(a\nabla u, \nabla u) \\
&\leq 2\|\mathbf{q}\|_{\mathbf{V},a}^2 + 2(a\nabla u, \nabla u)_{0,\Omega} + 2(du, u) + 2(a\nabla u, \nabla u)_{0,\Omega} \\
&\leq 4\|(\mathbf{q}, u)\|_a^2 \ .
\end{aligned}$$

For $(\mathbf{v}, w) := (\mathbf{q} + a\nabla u, -2u)$, this can be used to obtain

$$\begin{aligned}
\frac{1}{2}\|(\mathbf{q}, u)\|_a \|(\mathbf{v}, w)\|_a &\leq \|u\|_{W_0,a}^2 + \|\mathbf{q}\|_{\mathbf{V},a}^2 \\
&= (a^{-1}\mathbf{q}, \mathbf{q})_{0,\Omega} + (a\nabla u, \nabla u)_{0,\Omega} + (du, u)_{0,\Omega} \\
&\leq (a^{-1}\mathbf{q}, \mathbf{q})_{0,\Omega} + (a\nabla u, \nabla u)_{0,\Omega} + (2du, u)_{0,\Omega} \\
&= \left| \mathcal{L}\big((\mathbf{q}, u), (\mathbf{v}, w)\big) \right| \ .
\end{aligned}$$

Hence we can apply the generalized Lax-Milgram Theorem [7] to conclude that the continuous and linear operator $\mathcal{L}$ from $\mathbf{V} \times W$ to its dual space is bijective with the bound

$$\|(\mathbf{q}, u)\|_a \leq C_P \|(l_1, l_2)\| ,$$

where $C_P = 2$. Using the equivalent problem independent norm $\|\cdot\|$ defined by

$$\|(\mathbf{q}, u)\|_{\mathbf{V} \times W_0} := \|\mathbf{q}\|_{0,\Omega} + \|u\|_{1,\Omega}$$

the last estimate reads

$$\|(\mathbf{q}, u)\|_{\mathbf{V} \times W_0} \leq C_P \|(l_1, l_2)\|_{\mathbf{V}^* \times W_0^*} ,$$

with $C_P = 2 \max\{\|a\|_\infty, \|a^{-1}\|_\infty (C_{PF}^2 + 1)\}$, where C_{PF} is the constant in the Poincaré-Friedrichs inequality

$$\|w\|_{0,\Omega} \leq C_{PF} \|\nabla w\|_{0,\Omega} .$$

$\square$

2.1.3 Dual mixed formulation

On the other hand, the dual mixed formulation of (2.1) requires the computation of $(\mathbf{q}, u) \in \mathbf{V} \times W$, where $\mathbf{V} := H(div; \Omega)$, $W := L^2(\Omega)$, such that

$$a_D(\mathbf{q}, \mathbf{v}) + b_D(u, \mathbf{v}) = l_1(\mathbf{v}) \qquad \forall \mathbf{v} \in \mathbf{V} , \qquad (2.8a)$$
$$b_D(w, \mathbf{q}) - d_D(u, w) = l_2(w) \qquad \forall w \in W . \qquad (2.8b)$$

Here, the bilinear forms $a_D(\cdot, \cdot), b_D(\cdot, \cdot), d_D(\cdot, \cdot)$ are defined according to

$$a_D(\mathbf{q}, \mathbf{v}) := (a^{-1}\mathbf{q}, \mathbf{v})_{0,\Omega}, \qquad b_D(u, \mathbf{v}) := -(u, \nabla \cdot \mathbf{v})_{0,\Omega}, \qquad d_D(u, w) := (du, w)_{0,\Omega},$$

and the functionals $l_1 \in \mathbf{V}^*$ and $l_2 \in W^*$ are given by

$$l_1(\mathbf{v}) := -\langle \nu_\Gamma(\mathbf{q}), g_D \rangle_\Gamma, \qquad l_2(w) := (-f, w)_{0,\Omega} .$$

The operator $B : \mathbf{V} \to W^*$ defined by $\langle B\mathbf{v}, u \rangle_{W^*, W} = b_D(u, \mathbf{v})$ is surjective, in particular the range of B is closed in W^*. Together with the fact that we have $ker(B^*) = \{0\}$ for the adjoint B^* this implies that for any $(l_1, l_2) \in \mathbf{V}^* \times W^*$, problem (2.8) admits a unique solution $(\mathbf{q}, u) \in \mathbf{V} \times W$ (cf., e.g. Theorem 2.1 of [13]).

In order to exhibit the connection between both the primal and the dual formulation we prove the following remark inspired by the proof of the well-posedness of the dual mixed formulation for the Poisson equation given in [49]:

Remark 2.1.2

Suppose $f \in L^2(\Omega)$. For the solution u of the primal formulation (2.2) we set $\mathbf{q} := -a\nabla u$. Then $(\mathbf{q}, u)$, the solution $(\mathbf{q}_P, u_P)$ of the primal mixed formulation (2.4) and the solution $(\mathbf{q}_D, u_D)$ of the dual mixed formulation (2.8) coincide in the sense that $u = u_P = u_D \in H_0^1(\Omega)$ and $\mathbf{q} = \mathbf{q}_P = \mathbf{q}_D \in H(div; \Omega)$.

Proof. Clearly $-a\nabla u \in L^2(\Omega)^2$. Rearranging the terms in (2.2), we obtain

$$(\mathbf{q}, \nabla w)_{0,\Omega} = (du - f, w)_{0,\Omega} \qquad \forall w \in H_0^1(\Omega) \ .$$

Since $H_0^1(\Omega)$ is dense in $L^2(\Omega)$ the above variational equation implies that the weak divergence $\nabla \cdot \mathbf{q}$ exists, and the fact that $f \in L^2(\Omega)$ implies that $\nabla \cdot \mathbf{q} = du - f \in L^2(\Omega)$. This implies that for the solution u of (2.2) with $f \in L^2(\Omega)$ we actually have $\mathbf{q} = -a\nabla u \in H(div; \Omega)$. The same argument, applied to equation (2.4a), implies that $\mathbf{q}_P \in H(div; \Omega)$.

We know, that the three formulations (2.2), (2.4) and (2.8) admit unique solutions. Hence, it remains to check if $(-a\nabla u, u)$ and $(-a\nabla u_P, u_P)$ satisfy (2.8). Since $\mathbf{q} = -a\nabla u \in H(div; \Omega)$ we can use integration by parts to get

$$(a^{-1}\mathbf{q}, \mathbf{v}) = -(\nabla u, \mathbf{v}) = (u, \nabla \cdot \mathbf{v}) - \langle g_D, \mathbf{v} \cdot \mathbf{n} \rangle_\Gamma \ \forall \mathbf{v} \in H(div; \Omega) \qquad (2.9)$$

which is (2.8a). For the primal mixed method we know from equation (2.4a) that $(a^{-1}\mathbf{q}_P + \nabla u_P, \mathbf{v}) = 0 \ \forall \mathbf{v} \in L^2(\Omega)$, which is true, in particular, for all $\mathbf{v} \in H(div; \Omega) \subset L^2(\Omega)$. So the same argument as (2.9) implies that $(\mathbf{q}_P, u_P)$ satisfies (2.8a).

On the other hand we know from (2.2) – using integration by parts – that

$$(\nabla \cdot \mathbf{q}, w)_{0,\Omega} + (du, w)_{0,\Omega} - (f, w)_{0,\Omega} = 0 \qquad \forall w \in H_0^1(\Omega) \ .$$

Since $H_0^1(\Omega)$ is dense in $L^2(\Omega)$ we conclude directly that

$$(\nabla \cdot \mathbf{q} + du, w) = (f, w) \qquad \forall w \in L^2(\Omega) \ ,$$

which is (2.8b). Again the same argument applied to equation (2.4b) implies that $(\mathbf{q}_P, u_P)$ satisfies (2.8b). $\qquad \square$

2.1.4 Towards discretization

A popular and elegant method to compute a numerical approximation to the solution of the primal variational formulation is to restrict (2.2) to a finite dimensional subspace $W_h \in H_0^1(\Omega)$ and compute a solution $u_h \in W_h$ to the variational equation. Therefore we can divide Ω into a mesh of simplices T, choose $W_h(T)$ to be the space of polynomials of some finite degree on each simplex T and guarantee H^1-conformity of the space W_h by requiring continuity across interior faces of the mesh. However, as we only guarantee H^1-conformity of u we have the drawback that $\mathbf{q}_h = -a\nabla_h u_h$ in opposite to the exact flux $\mathbf{q} = -a\nabla u$ is actually not in $H(div; \Omega)$. This problem is tackled when discretizing the dual mixed formulation (2.8). In this case one is interested in a discrete approximation $\mathbf{q}_h \approx \mathbf{q}$ that respects that $\mathbf{q} \in H(div; \Omega)$. This task is met by enforcing continuity of the normal component of $\mathbf{q}_h$ across interior edges of the triangulation.

This so called *jump condition* is the key to all hybridized methods considered in this thesis. In the case of hybrid mixed methods like the well known Hybridized Raviart-Thomas (H-RT) method or the Hybridized Brezzi-Douglas-Marini (H-BDM) method $H(div; \Omega)$-conformity is directly guaranteed while in the case of Hybridized Discontinuous Galerkin (H-DG) methods we only guarantee the single-valuedness of the so-called numerical flux $\hat{\mathbf{q}}_h$ which is an approximation to the trace of $\mathbf{q}$ on the skeleton of the triangulation. The methods all have in common that they can be built based on this *jump condition*.

2.2 Idea of hybridization

In order to get the main idea of hybridization we consider the solution $(-a\nabla u, u) \in \mathbf{V} \times W$ of the PDE in a suitable pair of function spaces. For any fixed triangulation $\mathcal{T}$ of Ω we define the trace of u on the interior edges $\mathcal{E}^o$ of the triangulation as

$$\lambda := u|_{\mathcal{E}^o}. \tag{2.10}$$

In order to combine λ with the Dirichlet boundary data g_D we introduce the following formal extensions

$$\lambda = \begin{cases} u|_{\mathcal{E}^o} & \text{on} \quad \mathcal{E}^o \\ 0 & \text{on} \quad \Gamma \end{cases} \text{ and } g = \begin{cases} 0 & \text{on} \quad \mathcal{E}^o \\ g_D & \text{on} \quad \Gamma \end{cases}. \tag{2.11}$$

Now we can identify the trace of the solution on the skeleton of the triangulation as $u = \lambda + g$. The solution of the restriction of the global partial differential equation to a single element T depends linearly on the load f, the boundary data g_D and λ, where λ is a Dirichlet boundary condition to the boundary value problem restricted to one element. We define so called *local solvers* $(\mathbf{Q}_m(\cdot), \mathbf{U}_m(\cdot))$ for the traces mapping from some subspace $M \subset L^2(\partial T)$ to $\mathbf{V} \times W$ by $(\mathbf{Q}_m(\cdot), \mathbf{U}_m(\cdot)) : M \to \mathbf{V} \times W; \; m \mapsto (\mathbf{q}_m, u_m)$ for solving the BVP with zero right hand side on each element separately.

$$
\begin{aligned}
a^{-1}\mathbf{q}_m + \nabla u_m &= 0, \quad \text{in } T \\
\nabla \cdot \mathbf{q}_m + d u_m &= 0 \quad \text{in } T, \\
u_m &= m \quad \text{on } \partial T,
\end{aligned}
\tag{2.12}
$$

Analogously we define a *local solver* for the load $(\mathbf{Q}_f(\cdot), \mathbf{U}_f(\cdot)) : \; L^2(\Omega) \to \mathbf{V} \times W; \; f \mapsto (\mathbf{q}_f, u_f)$ for solving the following problem with homogeneous boundary values and right hand side f on T:

$$
\begin{aligned}
a^{-1}\mathbf{q}_f + \nabla u_f &= 0, \quad \text{in } T \\
\nabla \cdot \mathbf{q}_f + d u_f &= f \quad \text{in } T, \\
u_f &= 0 \quad \text{on } \partial T,
\end{aligned}
\tag{2.13}
$$

We split the local problems into these two kinds of local solvers because we need them for different purposes. The first of them assigns $(\mathbf{q}_m, u_m)$ to the unknown trace λ while the second solver assigns $(\mathbf{q}_f, u_f)$ to a priori known data. If the jump condition

$$
[\![\mathbf{q}_\lambda + \mathbf{q}_g + \mathbf{q}_f]\!] = 0
\tag{2.14}
$$

is satisfied the linearity of our problem gives the solution of the original boundary value problem as $(\mathbf{q}, u) = (\mathbf{Q}_m \lambda + \mathbf{Q}_m g + \mathbf{Q}_f f, \mathbf{U}_m \lambda + \mathbf{U}_m g + \mathbf{U}_f f)$.

The basis of the framework of hybridized methods as it has been introduced in [25] is a suitable version of the above jump condition. Additionally one needs suitable discrete local solvers. Therefore we choose solvers for well known existing finite element methods. So for all hybridized methods one first has to calculate an approximation to λ using a discrete version of the jump condition before achieving approximations to $\mathbf{q}$ and u by solving local problems. The title *Hybrid Method* is motivated by the fact that the methods provide both an approximation to the solution inside the element as well as an approximation to the trace of the solution on the boundaries of the elements.

For numerical computations we choose a shape regular family of geometrically conforming simplicial triangulations $(\mathcal{T})_\mathbb{H}$ of Ω. For each $\mathcal{T} \in (\mathcal{T})_\mathbb{H}$ we denote by $\mathcal{N}$ the set of its vertices, by $\mathcal{E}^0$ the set of its interior edges, and by $\mathcal{E}^\partial$ the set of the edges located on the boundary Γ. The set of all edges is denoted by $\mathcal{E} := \mathcal{E}^0 \cup \mathcal{E}^\partial$. Moreover, for $D \subset \Omega$ the sets $\mathcal{N}^D$ and $\mathcal{E}^D$ stand for the sets of the vertices and the edges of $\bar{D}$. For $T \in \mathcal{T}$ and $E \in \mathcal{E}$, we further denote by h_T and h_E the diameter of T and the length of E, and by ω_T and ω_E

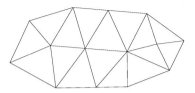

Figure 2.1: Triangulation $\mathcal{T}$ of a domain $\Omega \subset \mathbb{R}^2$.

the sets

$$\omega_T := \bigcup \{T' \in \mathcal{T} | \mathcal{N}^{T'} \cap \mathcal{N}^T \neq \emptyset\} \ ,$$
$$\omega_E := \bigcup \{E' \in \mathcal{E} | \mathcal{N}^{E'} \cap \mathcal{N}^E \neq \emptyset\} \ .$$

We refer to $H^k(\mathcal{T}), k \in \mathbb{N}$ as the Hilbert space

$$H^k(\mathcal{T}) := \prod_{T \in \mathcal{T}} H^k(T) \supset H^k(\Omega) \ ,$$

equipped with the inner product

$$(v,w)_{k,\mathcal{T}} := \sum_{T \in \mathcal{T}} (v,w)_{k,T} \ , \ v,w \in H^k(\mathcal{T}) \ ,$$

and the associated norm $\|\cdot\|_{k,\mathcal{T}}$. In contrast, we denote for the Hilbert space $L^2(E)$, $E \in \mathcal{E}$ the inner product

$$\langle u,v \rangle_E := \int_E uv \, d\sigma \ , \ u,v \in L^2(E) \ .$$

Whereas we denote the corresponding space on $\mathcal{E}^i$ as $L^2(\mathcal{E}^i)$ equipped with the inner product $\langle \cdot, \cdot \rangle_{\mathcal{E}^i}$ defined as

$$\langle u,v \rangle_{\mathcal{E}} := \sum_{E \in \mathcal{E}} \langle u,v \rangle_E \ , \forall u,v \in L^2(\mathcal{E}^i) \ .$$

We denote the corresponding Hilbert space norms by $\|v\|_{L^2(E)} =: \|v\|_{0,E}$ and $\|v\|_{L^2(\mathcal{E}^i)} =: \|v\|_{0,\mathcal{E}^i}$ for any subset $\mathcal{E}^i \subset \mathcal{E}$.

Remark 2.2.1

We use the same notation

$$\langle u,v\rangle_E := \int_E uv \, d\sigma \, , \forall u \in H^{1/2}(E) \, , v \in H^{-1/2}(E)$$

to denote the dual paring between $H^{1/2}(E)$ and $H^{-1/2}(E)$. It automatically becomes clear by the spaces, of which u and v are chosen, if $\langle \cdot, \cdot \rangle_E$ denotes the inner product or the dual paring.

In the same spirit we use the following abbreviation

$$\langle u,v\rangle_{\partial\mathcal{T}} := \sum_{T\in\mathcal{T}} \langle u,v\rangle_{\partial T} \, ,$$

where we have to remark that here – in contrast to the definition of $\langle .,.\rangle_{\mathcal{E}}$ – on every interior face $E \subset \partial T^+ \cap \partial T^-$ the inner product is double valued with one branch belonging to T^+ and one branch belonging to T^-.

Figure 2.2: Detail of the graph of a function $u \in H^1(\mathcal{T})$

Generally speaking, functions in $u \in H^l(\mathcal{T})$ and $\mathbf{v} \in H^l(\mathcal{T})^2$ are not globally continuous (see 2.2) for any $l \geq 1$. So the integration by parts formula can only be applied element-wise.

$$(u,\nabla\cdot\mathbf{v})_{\mathcal{T}} + (\nabla u,\mathbf{v})_{\mathcal{T}} = \langle u,\mathbf{v}\cdot\mathbf{n}\rangle_{\partial\mathcal{T}}. \tag{2.15}$$

In order to rewrite the left hand side with respect to the sum across edges it is useful to introduce the concept of jumps and means first.

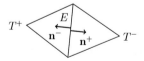

Figure 2.3: Elements T^+, T^- and their outward pointing normals on $E = \mathcal{T}^+ \cap \mathcal{T}^-$.

For any interior face $E = T^+ \cap T^-$ between to elements T^+ and T^- (see Figure 2.3) the jump of a function $u \in H^1(\mathcal{T})$ is defined by

$$[\![u]\!]_E := u^+ \mathbf{n}^+ + u^- \mathbf{n}^- , \qquad (2.16)$$

where u^+ denotes the trace of the function u restricted to T^+ on E and u^- denotes the trace of u restricted to T^- on E. Similarly, we define the jump of a vector field $\mathbf{q} \in H(div; \mathcal{T})$ across E as

$$[\![\mathbf{q}]\!]_E := \mathbf{q}^+ \cdot \mathbf{n}^+ + \mathbf{q}^- \cdot \mathbf{n}^- . \qquad (2.17)$$

On exterior faces E the jumps are defined by $[\![u]\!]_E := u\mathbf{n}$ and $[\![\mathbf{q}]\!]_E := \mathbf{q} \cdot \mathbf{n}$, respectively. Similarly we denote the arithmetic mean of a function $u \in H^1(\mathcal{T})$ and $\mathbf{q} \in H(div; \mathcal{T})$ by

$$\{u\}_E := \frac{1}{2}(u^+ + u^-) \qquad \text{and} \qquad \{\mathbf{q}\}_E := \frac{1}{2}(\mathbf{q}^+ + \mathbf{q}^-) . \qquad (2.18)$$

while on exterior faces E we set $\{u\}_E := u$ and $\{\mathbf{q}\}_E := \mathbf{q}$. Using this notation one can see by a simple rearrangement of terms that the following identity holds for any $u \in H^1(\mathcal{T})$ and $\mathbf{v} \in H(div; \mathcal{T})$:

$$\langle u, \mathbf{v} \cdot \mathbf{n} \rangle_{\partial \mathcal{T}} = \langle [\![u]\!], \{\mathbf{v}\} \rangle_{\mathcal{E}} + \langle \{u\}, [\![\mathbf{v}]\!] \rangle_{\mathcal{E}^0} . \qquad (2.19)$$

Chapter 3

Hybrid methods

3.1 General structure

In this chapter, we specify the problem on the discrete level. We show which conditions should be imposed on hybridized methods in order to get a unique numerical solution. We follow the classification of hybrid methods introduced in [25].

For each $T \in \mathcal{T}$ we denote the space of polynomial vectorfields of total degree at most k_1 by $\mathbf{V}(T) := \mathbb{P}_{k_1}(T)^2$ and similarly, for a scalar polynomial field of total degree at most k_2 we set $W(T) := \mathbb{P}_{k_2}(T)$. For global approximations, we define the product spaces

$$\mathbf{V}_h := \prod_{T \in \mathcal{T}} \mathbf{V}(T) \qquad \text{and} \qquad W_h := \prod_{T \in \mathcal{T}} W(T) \ .$$

The H^1-conforming finite element spaces are denoted by

$$W_{k_2}^c = \prod_{T \in \mathcal{T}} \mathbb{P}_{k_2}(T) \cap H_0^1(\Omega) \ .$$

Analogously, for discretization of the trace space we define a finite dimensional polynomial version $M(\partial T) \subset L^2(\partial T)$. The discrete traces are required to be single valued on interior edges. Hence we set

$$M_h := \prod_{T \in \mathcal{T}} M(\partial T) \cap L^2(\mathcal{E}) \ .$$

To be more precise, we approximate the trace of u on $\mathcal{E}$ by $m_h = \lambda_h + g_h \in M_h$, where $\lambda_h \in M_h^0 := \{\mu \in M_h : \mu|_{\partial\Omega} = 0\}$ and $g_h = I_{M_h} g$ is a suitable interpolation of the function g which is defined on $\partial\Omega$ into $\mathcal{E}^o$. In the case – when functions in M_h do not have to be continuous on $\mathcal{E}$ – $I_{M_h} g$ is usually defined as the extension of g by zero on the set of interior edges $\mathcal{E}^o$.

The following lemmas give relations between norms of functions in W_h and its traces on the skeleton of the triangulation.

Lemma 3.1.1

Let $v \in \prod_{T \in \mathcal{T}} \mathbb{P}_k(T)$ over a conforming nondegenerate triangulation $\mathcal{T}$. Then we have

$$\|\mathbf{n}_{\mathcal{E}}[\![v]\!]\|_{0,\mathcal{E}} \lesssim C_k \|h_{\mathcal{T}}^{-1/2} v\|_{0,\mathcal{T}} \ , \tag{3.1}$$

as well as

$$\|\{v\}\|_{0,\mathcal{E}} \lesssim C_k \|h_{\mathcal{T}}^{-1/2} v\|_{0,\mathcal{T}} \ , \tag{3.2}$$

where $C_k \le \sqrt{3(k+1)(k+2)} \le 2(k+1)$ and the hidden constants depend only on the local geometry of the triangulation.

Proof. The proof of this inequality is based on an inverse trace inequality on the reference element

$$\hat{T} = \{(s,t) \in \operatorname{convhull}((-1,-1),(1,-1),(-1,1))\}$$

with edge $\hat{E} := \operatorname{convhull}((-1,-1),(-1,1))$. There is a basis of orthogonal polynomials for $\mathbb{P}_k(\hat{T})$ given by

$$\psi_{i,j}(s,t) := \alpha_{ij} P_i^{(0,0)} \big(\frac{2(s+1)}{1-t} - 1 \big)(1-t)^i \ P_j^{(2i+1,0)}(t) \ ,$$

where $P_k^{(\beta,\gamma)}$ denote the classical Jacobi polynomials of order k, see [46] for details. For the restrictions of these polynomials to the edge $\hat{E}$, where $t = -1$, we have

$$\psi_{i,j}(s,-1) := c_{ij} P_i^{(0,0)}(s) \ .$$

The classical Jacobi polynomials for $(\beta,\gamma) = (0,0)$ coincide with the well known Legendre polynomials which are orthogonal on $\hat{E}$. Therefore, we have the estimate

$$\|v\|_{0,E}^2 = \|\sum_{i,j=0}^{k} v_i \psi_{i,j}\|_{0,\hat{E}}^2 = \sum_{i,j=0}^{k} |v_i|^2 \|\psi_{i,j}\|_{0,\hat{E}}^2 \le C_k^2 \sum_{i=1}^{k+1} |v_i|^2 \|\psi_{i,j}\|_{0,\hat{T}}^2$$

with a constant C_k depending on k. In [59] it has been shown that $C_k \le \sqrt{\frac{(k+1)(k+2)}{2}}$ which can be roughly estimated by $C_k \le (k+1)$. Using this estimate and a simple scaling argument we conclude

$$\|\mathbf{n}_{\mathcal{E}}[\![v]\!]\|_{0,\mathcal{E}_h}^2 = \sum_{E\in\mathcal{E}_h} \|v^+ - v^-\|_{0,E}^2 \le 2\sum_{E\in\mathcal{E}_h} (\|v^+\|_{0,E}^2 + \|v^-\|_{0,E}^2)$$

$$\lesssim 2C_k^2 \sum_{E\in\mathcal{E}_h} h_E^{-1}(\|v^+\|_{0,T^+}^2 + \|v^-\|_{0,T^-}^2) \lesssim 6C_k^2 \sum_{T\in\mathcal{T}} h_T^{-1}\|v\|_{0,T}^2 = 6C_k^2\|h_{\mathcal{T}}^{-1/2}v\|_{0,\mathcal{T}}^2 \ .$$

Analogously we proceed for $\{v\}$. □

Using this estimate, the Cauchy-Schwarz and Young's inequality, we easily get the following bounds.

Corollary 3.1.2
For any two functions v, $w \in W_h = \prod_{T\in\mathcal{T}} \mathbb{P}_k(T)$ over a conforming nondegenerate triangulation $\mathcal{T}$ and for any fixed $\tilde{\kappa} > 0$ there is a constant C that only depends on the local geometry of the triangulation such that

$$|\langle [\![v]\!], \{a\nabla w\}\rangle_{\mathcal{E}_h}| \le \frac{C(k+1)}{\tilde{\kappa}}(a\nabla w, \nabla w) + \tilde{\kappa}\langle h^{-1}[\![v]\!], [\![v]\!]\rangle_{\mathcal{E}_h} \tag{3.3}$$

and

$$|\langle [\![v]\!], [\![a\nabla w]\!]\rangle_{\mathcal{E}_h}| \le \frac{C(k+1)}{\tilde{\kappa}}(a\nabla w, \nabla w) + \tilde{\kappa}\langle h^{-1}[\![v]\!], [\![v]\!]\rangle_{\mathcal{E}_h} \ . \tag{3.4}$$

To motivate the general form of the local solvers we start from the dual mixed variational formulation of the local solver (2.12) with respect to the finite dimensional spaces $W(T), \mathbf{V}(T)$ and $M(\partial T)$. In order to include the hybridization of discontinuous Galerkin methods into the general framework we use integration by parts in the second equation and replace the normal trace of $\mathbf{Q}_m m$ on ∂T by a newly defined variable $\hat{\mathbf{q}}_m$, called *numerical flux*, supported only on the edges of the triangulation.

We thus define the discrete local solver

$$(\mathbf{Q}(\cdot), \mathbf{U}(\cdot)) : \ M(\partial T) \to \mathbf{V}(T) \times W(T); \ m \mapsto (\mathbf{q}_m, u_m)$$

by the unique solution of

$$(a^{-1}\mathbf{q}_m, \mathbf{v})_T - (u_m, \nabla\cdot\mathbf{v})_T = -\langle m, \mathbf{v}\cdot\mathbf{n}\rangle_{\partial T} \qquad \forall \mathbf{v} \in \mathbf{V}(T) \ , \tag{3.5a}$$

$$-(\mathbf{q}_m, \nabla w)_T + (du_m, w)_T + \langle w, \hat{\mathbf{q}}_m\cdot\mathbf{n}\rangle_{\partial T} = 0 \qquad \forall w \in W(T) \ . \tag{3.5b}$$

The definition of $\hat{\mathbf{q}}_m$ is essential in the framework of hybridization. It depends linearly on m, $\mathbf{Q}m$ and $\mathbf{U}m$. The precise form of $\hat{\mathbf{q}}_m$ depends on the actual hybridized method under consideration. It makes the major difference between the individual methods.

Equivalently, for the load $f \in L^2(\Omega)$ we define the solver

$$(\mathbf{Q}(\cdot), \mathbf{U}(\cdot)) : \ L^2(T) \to \mathbf{V}(T) \times W(T); \ f \mapsto (\mathbf{q}_f, u_f) \ .$$

The system (2.13) is discretized by

$$(a^{-1}\mathbf{q}_f, \mathbf{v})_T - (u_f, \nabla\cdot\mathbf{v})_T = 0 \qquad \forall \mathbf{v} \in \mathbf{V}(T) \ , \qquad (3.6\text{a})$$

$$-(\mathbf{q}_f, \nabla w)_T + (du_f, w)_T + \langle w, \hat{\mathbf{q}}_f\cdot\mathbf{n}\rangle_{\partial T} = (f, w)_T \qquad \forall w \in W(T), \qquad (3.6\text{b})$$

where again the dependence of $\hat{\mathbf{q}}_f$ on $f, \mathbf{Q}f$ and $\mathbf{U}f$ differs from method to method.

In order to keep notation short, we denote both solvers by the same symbol $(\mathbf{Q}(\cdot), \mathbf{U}(\cdot))$, although they are different maps. The difference always becomes clear from the function that is mapped. We choose the first method, if the operator is applied to a function in $M(\partial T)$, while we choose the second method, if it is applied to a function in $L^2(T)$.

The concept of local solvers leads directly to the first assumption we need in order to get a well defined hybridized method.

Assumption 3.1.3
The spaces $\mathbf{V}(T)$, $W(T), M(\partial T)$ and the numerical fluxes $\hat{\mathbf{q}}_m$ and $\hat{\mathbf{q}}_f$ are chosen such that for every $m \in M(\partial T)$ and $f \in L^2(T)$ the local solvers (3.5) and (3.6) admit unique solutions.

Often it is useful to write the method in a more compact form. Therefore, we set $\mathbf{q}_h := \mathbf{q}_m + \mathbf{q}_f$, $u_h := u_m + u_f$ and $\hat{\mathbf{q}}_h := \hat{\mathbf{q}}_m + \hat{\mathbf{q}}_f$. On one hand $\hat{\mathbf{q}}_h$ depends linearly on $\mathbf{q}_h$, u_h and $m_h = \lambda_h + g_h$. On the other hand, $\mathbf{q}_h$ and u_h are the images of m_h and f under a linear mapping. So we may express $\hat{\mathbf{q}}_h$ as the image of g_h, λ_h and f under a linear mapping by $\hat{\mathbf{q}}_h = \hat{\mathbf{Q}}f + \hat{\mathbf{Q}}\lambda_h + \hat{\mathbf{Q}}g_h$.

By now we have discretized only the local solvers. In order to get a useful method to approximate the solution of the boundary value problem in the entire domain we need to couple the local problems globally. Global coupling of the numerical approximation of $(u, \mathbf{q})$ is realized by the following discretization of the *jump condition* for the numerical fluxes:

$$\langle [\![\hat{\mathbf{q}}_h]\!], \mu\rangle_{\mathcal{E}} = \langle [\![\hat{\mathbf{Q}}f + \hat{\mathbf{Q}}\lambda_h + \hat{\mathbf{Q}}g_h]\!], \mu\rangle_{\mathcal{E}} = 0 \qquad \forall \mu \in M_h^0 \ . \qquad (3.7)$$

We note that this general form of a hybridized method split into the local solvers and the global coupling through the discrete jump condition can be summarized in the following equivalent formulation. Find $(u_h, \mathbf{q}_h, \lambda_h) \in W_h \times \mathbf{V}_h \times M_h^0$ satisfying

$$(a^{-1}\mathbf{q}_h, \mathbf{v})_{\mathcal{T}} - (u_h, \nabla \cdot \mathbf{v})_{\mathcal{T}} + \langle \lambda_h, \mathbf{v} \cdot \mathbf{n} \rangle_{\partial \mathcal{T} \setminus \partial \Omega} = -\langle g_h, \mathbf{v} \cdot \mathbf{n} \rangle_{\partial \Omega} \qquad \forall \mathbf{v} \in \mathbf{V}_h \ , \qquad (3.8a)$$

$$-(\mathbf{q}_h, \nabla w)_{\mathcal{T}} + (du_h, w)_{\mathcal{T}} + \langle w, \hat{\mathbf{q}}_h \cdot \mathbf{n} \rangle_{\partial \mathcal{T}} = (f, w)_{\mathcal{T}} \qquad \forall w \in W_h \ , \qquad (3.8b)$$

$$\langle [\![\hat{\mathbf{q}}_h]\!], \mu \rangle_{\mathcal{E}} = 0 \qquad \forall \mu \in M_h^0 \ . \qquad (3.8c)$$

Although this formulation does not reflect the structure of local problems being coupled by the discrete jump condition, it will be used frequently later on for the development of individual methods and the error analysis.

In order to show how the discrete *jump condition* is used to compute the value of λ_h we state the following theorem.

Theorem 3.1.4

The discrete jump condition (3.7) *is satisfied if and only if the linear system*

$$a_h(\lambda_h, \mu) = b_h(\mu) \qquad \forall \mu \in M_h^0 \ , \qquad (3.9)$$

admits a unique solution, where

$$a_h(\eta, \mu) = (a^{-1}\mathbf{Q}\eta, \mathbf{Q}\mu)_{\mathcal{T}} + (d\mathbf{U}\eta, \mathbf{U}\mu)_{\mathcal{T}} + \langle 1, [\![(\mathbf{U}\mu - \mu)(\hat{\mathbf{Q}}\eta - \mathbf{Q}\eta)]\!] \rangle_{\mathcal{E}} \ ,$$

$$\begin{aligned} b_h(\mu) = &\langle g_h, [\![\hat{\mathbf{Q}}\mu]\!] \rangle_{\mathcal{E}} + (f, \mathbf{U}\mu)_{\mathcal{T}} \\ &- \langle 1, [\![(\mathbf{U}\mu - \mu)(\hat{\mathbf{Q}}f - \mathbf{Q}f)]\!] \rangle_{\mathcal{E}} \\ &+ \langle 1, [\![(\mathbf{U}f)(\hat{\mathbf{Q}}\mu - \mathbf{Q}\mu)]\!] \rangle_{\mathcal{E}} \\ &- \langle 1, [\![(\mathbf{U}\mu - \mu)(\hat{\mathbf{Q}}g_h - \mathbf{Q}g_h)]\!] \rangle_{\mathcal{E}} \\ &+ \langle 1, [\![(\mathbf{U}g_h - g_h)(\hat{\mathbf{Q}}\mu - \mathbf{Q}\mu)]\!] \rangle_{\mathcal{E}} \ . \end{aligned}$$

This equivalence can be proved by rewriting (3.7) as

$$-\langle \mu, [\![\hat{\mathbf{Q}}\lambda_h]\!] \rangle_{\mathcal{E}} = \langle \mu, [\![\hat{\mathbf{Q}}g_h + \hat{\mathbf{Q}}f]\!] \rangle_{\mathcal{E}} \ ,$$

defining

$$a_h(\lambda_h, \mu) := -\langle \mu, [\![\hat{\mathbf{Q}}\lambda_h]\!] \rangle_{\mathcal{E}} \ , \qquad \text{and} \qquad b_h(\mu) := \langle \mu, [\![\hat{\mathbf{Q}}g_h + \hat{\mathbf{Q}}f]\!] \rangle_{\mathcal{E}} \ , \qquad (3.10)$$

using the characterizing equations of the local solvers and performing some algebraic manipulations using integration by parts. For details of the proof we refer to [25, Theorem 2.1].

Equation (3.9) represents the global system of the hybridized methods. Using the local solvers that are applied in the global solver it serves to compute the approximation λ_h to the trace of the solution of the boundary value problem on the edges of the triangulation. Having this trace at hand we can locally compute a global approximation to $(\mathbf{q}, u)$ – the solution of the boundary value problem – using local solvers (3.5) and (3.6).

We still need to justify the well-posedness of the general hybridized method. In addition to 3.1.3 we state the following conditions on the local solvers due to [25].

Assumption 3.1.5

The local solvers and the numerical fluxes in (3.5) and (3.6) are such that, for every $T \in \mathcal{T}$ it holds:

$$-\langle \mu, \hat{\mathbf{Q}}\mu \cdot \mathbf{n} \rangle_{\partial T} \geq 0 \quad \forall \mu \in M_h \; . \tag{3.11a}$$

Furthermore, there is an auxiliary function space $M^\sharp(\partial T)$ containing the set $\{m : m|_E \in \mathbb{P}_0(E) \; \forall E \in \mathcal{E}^o \cap \partial T\}$ such that

$$\text{for any } \mu \in M_h \text{ with } \langle \mu, \hat{\mathbf{Q}}\mu \cdot \mathbf{n} \rangle_{\partial T} = 0 \text{ there is a constant } C(T) : P_{\sharp,T}\mu = C(T) \; , \tag{3.11b}$$

where $P_{\sharp,T}$ denotes the $L^2(\partial T)$-orthogonal projection onto $M^\sharp(\partial T)$.

By definition of the projection this constant is the number $C(T)$ satisfying the equation $\langle C(T) - \mu, m \rangle_{\partial T} = 0$ for all $m \in M^\sharp(\partial T)$.

Setting $m := \mu, \mathbf{v} := \mathbf{q}_\mu$ and $w := u_\mu$ in (3.5) we see

$$-\langle \mu, \hat{\mathbf{Q}}\mu \cdot \mathbf{n} \rangle_{\partial T} = (a^{-1}\mathbf{q}_\mu, \mathbf{q}_\mu)_T + (du_\mu, u_\mu)_T + \langle (\hat{\mathbf{q}}_\mu - \mathbf{q}_\mu) \cdot \mathbf{n}, u_\mu - \mu \rangle_{\partial T} \; . \tag{3.12}$$

Hence (3.11a) can be interpreted as the positive semi-definiteness of the global solver restricted to an element T.

Assumption 3.1.6

Let $E = T^+ \cap T^-$ be an interior edge. For all $\mu \in M_h$ we assume that either $\mu|_E = P_{\sharp,T^+}\mu$ or $\mu|_E = P_{\sharp,T^-}\mu$.

We can guarantee the unique solvability of any hybridized finite element method satisfying these general assumptions. This is the statement of the following theorem due to [25].

Theorem 3.1.7

Suppose that Assumptions 3.1.3 , 3.1.5 and 3.1.6 are satisfied. Then there is a unique solution to the global system (3.9).

For completeness we provide a proof of this general result.

Proof. Under Assumption 3.1.3, for a given $\lambda \in M_h^0$ we have a unique local solution $\hat{\mathbf{Q}}\lambda$

on each element $T \in \mathcal{T}$. From the definition in (3.10), the bilinear form $a_h(\lambda_h, \mu_h)$ is well defined for all λ_h, $\mu_h \in M_h^0$. Since the number of equations equals the number of the unknowns of the linear system, uniqueness of a solution is equivalent to its existence. We establish uniqueness by showing the following implication. If $a_h(\mu, \mu) = 0$ for any $\mu \in M_h^0$, then $\mu = 0$. Therefore, we consider

$$0 = a_h(\mu, \mu) = -\langle \mu, [\![\hat{\mathbf{Q}}\mu]\!] \rangle_\varepsilon = -\sum_{T \in \mathcal{T}} \langle \mu, \hat{\mathbf{Q}}\mu \cdot \mathbf{n} \rangle_{\partial T} \ .$$

Each of the terms on the right-hand side of the equation above is nonnegative due to Assumption 3.1.5. So we conclude

$$\langle \mu, \hat{\mathbf{Q}}\mu \cdot \mathbf{n} \rangle_{\partial T} = 0 \ .$$

Now, let $E = T^+ \cap T^-$ be an interior edge and choose $m = \chi_E \in M^\sharp(\partial T)$ to be the indicator function of E. With this choice of m, the second part of Assumption 3.1.5 states that the L_2 projections $P_{\sharp, T^\pm} \mu$ satisfying $\langle P_{\sharp, T^\pm} \mu - \mu, 1 \rangle_E = 0$ equal constants $C(T^\pm)$ on each element. This implies

$$C(T^+) = P_{\sharp, T^+} \mu = \frac{1}{|E|} \langle 1, \mu \rangle_E = P_{\sharp, T^-} \mu = C(T^-) \ .$$

Using Assumption 3.1.6, we conclude that $\mu = P_{\sharp, T^-} \mu = P_{\sharp, T^+} \mu$. Hence, μ is equal to the same element-wise constant on each edge of the two adjacent elements of each interior edge. In particular, μ admits the same value on the boundary of an element at $\partial \Omega$. Since $\mu|_{\partial \Omega} = 0$, we have $\mu = 0 \in M_h^0$. $\qquad\square$

We note that there is no restriction on the shape of the individual elements. For example, we may build methods based on simplices or rectangles. Since global coupling of the method is only achieved through the approximate trace λ_h on the edges of the triangulation, different element geometries may be used on different parts of the triangulation. Furthermore, this general approach also includes the possibility of building methods on nonconforming triangulations or methods with varying polynomial degree. But it has to be mentioned that by now we only have achieved unique solvability of the abstract method but no result for the quality of the approximation. In the following chapters, we are going to analyze different kinds of methods while we restrict ourselves to conforming simplicial triangulations and a fixed but still arbitrary polynomial degree.

3.2   Hybridized mixed methods

To demonstrate that the just introduced characterization of hybridized methods makes sense we check whether the well known hybrid versions of the mixed methods due to Raviart and Thomas and those due to Brezzi, Douglas and Marini can be recovered within this unifying framework. This is the topic of the current section.

3.2.1   Raviart-Thomas method

For $\mathbf{x} \in \mathbb{R}^2$ we choose

$$\mathbf{V}(T) := RT_k(T) = \mathbb{P}_k(T)^2 + \mathbf{x}\,\mathbb{P}_k(T) \ , \tag{3.13a}$$

$$W(T) := \mathbb{P}_k(T) \ , \tag{3.13b}$$

$$M(\partial T) = \mathbb{P}_k(\partial T) := \prod_{E \in \partial T} \mathbb{P}_k(E) \ , \tag{3.13c}$$

where $k \in \mathbb{N}_0$. These are the local polynomial spaces of the classical Raviart-Thomas method that has been introduced in [49]. Since $\mathbf{x} \cdot \mathbf{n}$ is constant for all $\mathbf{x}$ on E, where $\mathbf{n}$ is the unit normal on E, we see immediately that for any $\mathbf{v} \in \mathbf{V}(T)$ we have $\mathbf{v} \cdot \mathbf{n} \in \mathbb{P}_k(E)$. An easy exercise in elementary calculus shows the important fact that the divergence map $\nabla \cdot : \mathbf{V}(T) \to W(T)$ is onto.

The following projection operator is associated to the Raviart-Thomas spaces.

Lemma 3.2.1

There is a projection operator $\Pi_k^{RT} : \mathcal{H}(div; T) \to RT_k(T)$ *associated to the Raviart-Thomas space* $RT_k(T)$ *defined by the function* $\Pi_k^{RT}\mathbf{v} \in RT_k(T)$ *which satisfies*

$$l_E^{RT}(\Pi_k^{RT}\mathbf{v} - \mathbf{v}) := \langle \Pi_k^{RT}\mathbf{v} - \mathbf{v}, p_k \cdot \mathbf{n} \rangle_E = 0 \qquad \forall p_k \in \mathbb{P}_k(E) \ \forall E \subset \partial T \ , \tag{3.14a}$$

$$l_T^{RT}(\Pi_k^{RT}\mathbf{v} - \mathbf{v}) := (\Pi_k^{RT}\mathbf{v} - \mathbf{v}, \mathbf{p}_{k-1})_T = 0 \qquad \forall \mathbf{p}_{k-1} \in \mathbb{P}_{k-1}(T)^2 \ , \tag{3.14b}$$

where $\mathcal{H}(div; T) := \{\mathbf{q} \in H(div; T); \mathbf{q} \cdot \mathbf{n}_E \in L^2(E) \ \forall E \subset \partial T\}$ *is a strict subspace of* $H(div; T)$.

$\Pi_k^{RT}\mathbf{v} = \mathbf{v}_h$ *satisfies the following approximation property*

$$\|\Pi_k^{RT}\mathbf{v} - \mathbf{v}\|_{0,T} \leq Ch^l|\mathbf{v}|_{l,T} \qquad \forall \mathbf{v} \in H^l(T)^2, \ 1 \leq l \leq k+1, \tag{3.15a}$$

$$\|\nabla \cdot (\Pi_k^{RT}\mathbf{v} - \mathbf{v})\|_{0,T} \leq Ch^l|\nabla \cdot \mathbf{v}|_{l,T} \qquad \forall \mathbf{v} \in H^1(T)^2 \ with \tag{3.15b}$$

$$\nabla \cdot \mathbf{v} \in H^l(T), \ 0 \leq l \leq k+1.$$

Proof. Since the number of characterizing equations is equal to $(k+1)(k+3) = dimRT_k$, in order to show that Π_k^{RT} is well defined by (3.14), it suffices to demonstrate that $\mathbf{v}_h = 0$ is the unique projection of $\mathbf{v} = 0$. This is easily concluded after observing that $(\nabla \cdot \mathbf{v}_h, p_k)_T = (\mathbf{v}_h, \nabla \cdot p_k)_T + \langle p_k, \mathbf{v}_h \cdot \mathbf{n} \rangle_{\partial T} \; \forall p_k \in \mathbb{P}_k(T)$. The approximation properties are established by scaling arguments. $\square$

More detailed proofs can be found, for example, in Section 6 of [52].

Remark 3.2.2
For the global Raviart-Thomas mixed method the functionals l_T^{RT} and l_E^{RT} of (3.14) are commonly used to define the degrees of freedom, since it is easy to ensure $H(div; \Omega)$-conformity of the global discrete space $RT_k(\Omega) := \prod_{T \in \mathcal{T}} RT_k(T) \cap H(div; \Omega)$. In the context of hybridized methods, however, conformity is achieved via the discrete *jump condition* (3.7) where the numerical fluxes and the test space are chosen by

$$\hat{\mathbf{Q}}m := \mathbf{Q}m \; , \quad \hat{\mathbf{Q}}f := \mathbf{Q}f \; , \tag{3.16a}$$

$$M_h = \mathbb{P}_k(\mathcal{E}) := \prod_{E \in \mathcal{E}} \mathbb{P}_k(E) \; . \tag{3.16b}$$

Since on all edges E of ∂T we have $\mathbf{v} \cdot \mathbf{n}_E \in \mathbb{P}_k(E)$ for all $\mathbf{v} \in RT_k(T)$, the *discrete jump condition* (3.7) implies that $[\![\mathbf{q}_h]\!]_E = 0$. We conclude that the flux $\mathbf{q}_h$ is actually an element of $H(div; \Omega)$.

The following lemma assures that Assumption 3.1.3 is satisfied.

Lemma 3.2.3
With the choices of (3.13) the local solvers (3.5) and (3.6) are well-defined .

Proof. Since the map is linear and the individual terms in (3.5) and (3.6) are well defined, it suffices to show that $(\mathbf{q}_m, u_m) = 0$ is the unique solution of the homogeneous system

$$(a^{-1}\mathbf{q}_m, \mathbf{v})_T - (u_m, \nabla \cdot \mathbf{v})_T = 0 \qquad \forall \mathbf{v} \in \mathbf{V}(T) \; ,$$
$$(\nabla \cdot \mathbf{q}_m, w)_T + (du_m, w)_T = 0 \qquad \forall w \in W(T),$$

which is the system we get with $\hat{\mathbf{q}}_m = \mathbf{q}_m$ after using integration by parts in the second equation. Keeping in mind that $\nabla \cdot \mathbf{V}(T) = W(T)$ we conclude that $(\mathbf{q}_m, u_m) = 0$ using the same arguments as in the proof for the uniqueness of the solution of the dual mixed formulation (2.8). $\square$

Lemma 3.2.4
Assumptions 3.1.5 and 3.1.6 are satisfied as well.

Proof. The first part of Assumption 3.1.5 is obviously satisfied, since $\hat{\mathbf{q}}_\mu = \mathbf{q}_\mu$. To accept the second part as well, we define $M^\sharp(\partial T) := M(\partial T)$ and start from

$$0 = -\langle \mu, \mathbf{q}_\mu \rangle_{\partial T} = (a^{-1}\mathbf{q}_\mu, \mathbf{q}_\mu)_T + (du_\mu, u_\mu)_T .$$

From this equation we conclude that $\mathbf{q}_\mu = 0$. Using integration by parts, (3.5a) reduces to

$$(\nabla u_\mu, \mathbf{v})_T - \langle u_\mu - \mu, \mathbf{v}\cdot\mathbf{n} \rangle_{\partial T} = 0 . \tag{3.17}$$

Since the Raviart-Thomas projection Π_k^{RT} (3.14) is well defined, we see that there exists a $\mathbf{v} \in \mathbf{V}(T)$ satisfying

$$(\mathbf{p}_{k-1}, \mathbf{v})_T = (\nabla u_\mu, \mathbf{p}_{k-1})_T \qquad\qquad \forall \mathbf{p}_{k-1} \in \mathbb{P}_{k-1}(T)^2 , \tag{3.18a}$$

$$\langle p_k, \mathbf{v}\cdot\mathbf{n} \rangle_E = -\langle u_\mu - \mu, p_k \rangle_E \qquad\qquad \forall p_k \in \mathbb{P}_k(E) \, \forall E \in \partial T . \tag{3.18b}$$

Choosing $\mathbf{p}_{k-1} = \nabla u_\mu$ and $p_k = u_\mu - \mu$ and plugging this $\mathbf{v}$ into (3.17), we arrive at

$$(\nabla u_\mu, \nabla u_\mu)_T + \langle u_\mu - \mu, u_\mu - \mu \rangle_{\partial T} = 0 ,$$

from which we conclude that $u_\mu \in \mathbb{P}_0(T)$. This again implies that μ is constant on ∂T.

Assumption 3.1.6 is immediately satisfied, since for T^+ and T^- with $T^+ \cap T^- = E$ we have $M^\sharp(\partial T^+) \supset M_h \subset M^\sharp(\partial T^-)$. So $P_{\sharp,T^+}\mu = \mu = P_{\sharp,T^-}\mu$. $\qquad\square$

Comparing the solution $(\mathbf{q}_h, u_h)$ of the hybridized Raviart-Thomas method with the solution of the classical Raviart-Thomas discretization of (2.8) with approximation spaces $\mathbf{V}_h := RT_k(\Omega) \subset H(div; \Omega)$ and $W_h := \prod_{T \in \mathcal{T}} \mathbb{P}_k(T)$ introduced in [49] and reminding that both methods admit unique solutions, we observe that they coincide.

Using the quasi best approximation property of a conforming method satisfying the discrete inf-sup condition in the context of a $H(div; \Omega) \times L^2(\Omega)$-conforming finite element method we conclude that

$$\|\mathbf{q} - \mathbf{q}_h\|_{div} + \|u - u_h\|_0 \leq C(\min_{\mathbf{v}_h \in \mathbf{V}_h} \|\mathbf{q} - \mathbf{v}_h\|_{div} + \min_{w_h \in W_h} \|u - w_h\|_0) . \tag{3.19}$$

For more regular problems, due to the approximation properties of the Raviart-Thomas projection (3.15) we conclude the following result similar to the original result of [49]. The proof is a direct consequence of equations (3.15) and (3.19).

Lemma 3.2.5

Let the data of the partial differential equation (2.2) be sufficiently regular so that $u \in$

$H^{k+1}(\Omega)$, $\mathbf{q} \in H^{k+1}(\Omega)^2$ and $\nabla \cdot \mathbf{q} \in H^{k+1}(\Omega)$. Then, for the hybrid method with local solvers (3.13) and the discrete jump condition (3.7) with the numerical flux of the hybridized Raviart-Thomas method defined by (3.16), the following a priori estimate holds true:

$$\|\mathbf{q} - \mathbf{q}_h\|_{div} + \|u - u_h\|_0 \le Ch^{k+1}(|u|_{k+1} + |\mathbf{q}|_{k+1} + |\nabla \cdot \mathbf{q}|_{k+1}) \ .$$

3.2.2 Brezzi-Douglas-Marini method

For the BDM method we define the local spaces as

$$\mathbf{V}(T) := BDM(T) = \mathbb{P}_k(T)^2, \tag{3.20a}$$

$$W(T) := \mathbb{P}_{k-1}(T) \ , \tag{3.20b}$$

$$M(\partial T) = \mathbb{P}_k(\partial T) := \prod_{E \in \partial T} \mathbb{P}_k(E) \ , \tag{3.20c}$$

with some $k \ge 1$. Again it is easy to see that the divergence operator $\nabla \cdot : \mathbf{V}(T) \to W(T)$ is onto and that $\mathbf{v} \cdot \mathbf{n}|_{\partial T} \in M(\partial T)$ for all $\mathbf{v} \in \mathbf{V}(T)$. By analogy to the hybridized Raviart-Thomas method of section 3.2.1 we choose the numerical fluxes and the space M_h according to

$$\hat{\mathbf{Q}}m := \mathbf{Q}m \ , \quad \hat{\mathbf{Q}}f := \mathbf{Q}f \ . \tag{3.21a}$$

$$M_h = \mathbb{P}_k(\mathcal{E}) := \prod_{E \in \mathcal{E}} \mathbb{P}_k(E) \ . \tag{3.21b}$$

For the BDM method, Assumptions 3.1.3, 3.1.5 and 3.1.6 can be justified using exactly the same arguments. The only difference is that the corresponding projection operator is defined in a slightly different manner (see, for example, [13]:§III.3.3). Instead of (3.18), there is a $\mathbf{v} \in \mathbf{V}(T)$ satisfying in particular

$$(\mathbf{v}, \nabla p_{k-1})_T = (\nabla u_\mu, \nabla p_{k-1})_T \qquad \forall p_{k-1} \in \mathbb{P}_{k-1}(T) \ , \tag{3.22a}$$

$$\langle p_k, \mathbf{v} \cdot \mathbf{n} \rangle_E = -\langle u_\mu - \mu, p_k \rangle_E \qquad \forall p_k \in \mathbb{P}_k(E) \, \forall E \in \partial T \ . \tag{3.22b}$$

These equations lead to the same results as for the Raviart-Thomas methods, if we choose $p_{k-1} := u_m$ and $p_k := u_m - m$. So the method admits a unique solution. Again it is convergent and for sufficiently regular problems we have the following a priori estimates.

Lemma 3.2.6

Let the data of the partial differential equation (2.2) be sufficiently regular such that $u \in H^k(\Omega)$, $\mathbf{q} \in H^k(\Omega)^2$ and $\nabla \cdot \mathbf{q} \in H^k(\Omega)$. Then, for the hybrid method with local solvers

	RT_k, $k \geq 0$	BDM_k, $k \geq 1$
$dim(\mathbf{V}_h)$	$(k+1)(k+3)$	$(k+1)(k+2)$
$dim(W_h)$	$\frac{1}{2}(k+1)(k+2)$	$\frac{1}{2}(k)(k+1)$
$\|\mathbf{q}_h - \mathbf{q}\|_{div} + \|u_h - u\|_0$	k+1	k
$\|\mathbf{q}_h - \mathbf{q}\|_0$	$k+1$	$k+1$
$\|u_h - u\|_0$	$k+1$	k
$\|u_h^* - u\|_0$	$k+2$	$k + 2 - \delta_{k,1}$

Table 3.1: Convergence order of the hybridized mixed methods

(3.20) *and the discrete jump condition* (3.7) *with numerical flux* (3.21) *of the hybridized BDM method the following a priori estimate holds true:*

$$\|\mathbf{q} - \mathbf{q}_h\|_{div} + \|u - u_h\|_0 \leq Ch^k(|u|_k + |\mathbf{q}|_k + |\nabla \cdot \mathbf{q}|_k) \ .$$

Hybridized mixed methods have been known and used since [4] adopting the ideas of [28]. The idea of hybridization, however, is different here. It is not the hybridization of a conforming approximation of the dual mixed formulation (2.8) but a hybrid method in the sense of Assumptions 3.1.3, 3.1.5 and 3.1.6 that is constructed in such a way that it is a conforming approximation of (2.8).

Remark 3.2.7

For sufficiently regular problems the additional information stored in the numerical trace λ_h can be exploited via local post-processing to derive an approximation u_h^* of the primal unknown u with even better approximation results. See, for example, [4], [12], [13], or [53] for alternative post-processing techniques.

We summarize the approximation results in Table 3.1. Here, k denotes the polynomial degree of the approximate trace space $\mathbb{P}_k(E)$ on a single face E. The size of the global linear system (3.9) is $kN_{\mathcal{E}^0}$ for both the $H - RT_k$ and the $H - BDM_k$ methods. We see, that one, who is interested in just a good approximation of $\mathbf{q}$ would prefere the H-BDM method, because both global systems are of equal size, but the local solvers of the BDM-method are cheaper.

3.3 Discontinuous Galerkin methods

In this section, we present the hybridized Discontinuous Galerkin methods as they have been introduced in [25] where the spaces and local solvers of the Discontinuous Galerkin (DG) methods are used. We start with the so-called Local DG (LDG) method.

3.3.1 Local DG method

To motivate the choice of the numerical fluxes, we start from a general form of the numerical fluxes of DG methods as they have been analyzed in [21].

$$\hat{\mathbf{q}}_h := \{\mathbf{q}_h\} - C_{11}\,[\![u_h]\!] - \mathbf{C_{12}}[\![\mathbf{q_h}]\!] \;, \tag{3.23a}$$

$$\hat{u}_h := \{u_h\} - \mathbf{C_{12}} \cdot [\![u_h]\!] - C_{22}[\![\mathbf{q_h}]\!] \;, \tag{3.23b}$$

where $C_{11} > 0$ and $C_{22} \geq 0$.

In our framework of hybridization, $\lambda_h = \hat{u}_h$ is the only globally coupled variable. So we have to guarantee that the numerical flux $\hat{\mathbf{q}}_h$ can be formulated purely local. That means it may only depend on the primal and dual variable from one single element and on the trace variable λ_h. Hence we rewrite (3.23b) as

$$\frac{1}{2}u_h^- - \mathbf{C_{12}} \cdot \mathbf{n}^-u_h^- - C_{22}\mathbf{n}^- \cdot \mathbf{q}_h^- = \hat{u}_h - \frac{1}{2}u_h^+ + \mathbf{C_{12}} \cdot \mathbf{n}^+u_h^+ + C_{22}\mathbf{n}^+ \cdot \mathbf{q}_h^+ \;.$$

For simplicity we set $\mathbf{C_{12}} := \beta \cdot \mathbf{n}^-$. We see that we get an element-wise expression for $\hat{\mathbf{q}}_h$ on T^+, if we insert this term into (3.23a) after choosing $C_{11} := \tau(\beta - 1/2)$ and $\tau C_{22} := \beta - 1/2$, namely

$$\hat{\mathbf{q}}_h = \mathbf{q}_h^+ + \tau(u_h^+ - \hat{u}_h)\mathbf{n}^+ \;,$$

$$\hat{\mathbf{q}}_h = \mathbf{q}_h^- + \tau(u_h^- - \hat{u}_h)\mathbf{n}^- \;.$$

The second equation is clearly true by a symmetry argument.

This motivates the following definition of the numerical flux for the hybridized local Discontinuous Galerkin method. For any of the three choices of local spaces

$$\mathbf{V}(T) := \mathbb{P}_k(T)^2 \;, \qquad W(T) := \mathbb{P}_{k-1}(T) \;, \qquad k \geq 1, \, \tau_T \geq 0 \text{ on } \partial T \text{ , or} \tag{3.24a}$$

$$\mathbf{V}(T) := \mathbb{P}_k(T)^2 \;, \qquad W(T) := \mathbb{P}_k(T) \;, \qquad k \geq 0, \, \tau_T > 0 \text{ on at least} \tag{3.24b}$$
$$\text{one face of } \partial T \text{, or}$$

$$\mathbf{V}(T) := \mathbb{P}_{k-1}(T)^2 \;, \qquad W(T) := \mathbb{P}_k(T) \;, \qquad k \geq 1, \, \tau_T > 0 \text{ on } \partial T \;, \tag{3.24c}$$

using the local solvers $\mathbf{Q}$ and $\mathbf{U}$ and the outward pointing normal $\mathbf{n}$, on each element

$T \in \mathcal{T}$

$$\hat{\mathbf{Q}}m := \mathbf{Q}m + \tau(\mathbf{U}m - m)\mathbf{n} \ , \quad \hat{\mathbf{Q}}f := \mathbf{Q}f + \tau(\mathbf{U}f)\mathbf{n} \ , \qquad (3.25\text{a})$$

$$M_h = \mathbb{P}_k(\mathcal{E}) := \prod_{E \in \mathcal{E}} \mathbb{P}_k(E) \ . \qquad (3.25\text{b})$$

Lemma 3.3.1
For any of the choices of the local spaces in (3.24) together with the numerical fluxes of (3.25), local solvers (3.5) and (3.6) are well defined.

Lemma 3.3.2
Assumptions 3.1.5 and 3.1.6 are satisfied as well.

A transparent proof of these two lemmas can be found in [25].

Consider the discrete jump condition (3.7):

$$0 = \langle [\![\hat{\mathbf{q}}_h]\!], \mu \rangle_{\mathcal{E}^0} \forall \mu \in M_h^0 \ .$$

Adding the images of the local maps $\hat{\mathbf{Q}}$ for m and for f on both sides of an interior face $E \in \mathcal{E}_h^0$, we have $\hat{\mathbf{q}}_h^\pm = \mathbf{q}_h^\pm + \tau^\pm(u_h^\pm - \lambda_h)\cdot\mathbf{n}$ with normal trace $\hat{\mathbf{q}}_h^\pm \cdot \mathbf{n}^\pm \in M_h^0$. Consequently, for any of the three choices in (3.24) we see directly that $[\![\hat{\mathbf{q}}_h]\!]_E = 0$ for all faces $E \in \mathcal{E}^0$. This equation can be written as

$$[\![\mathbf{q}_h]\!] + \tau^+ u_h^+ + \tau^- u_h^- - (\tau^+ + \tau^-)\lambda_h = 0 \ .$$

In the case $\tau^+ + \tau^- \neq 0$, we can solve it for λ_h to get

$$\lambda_h = \frac{\tau^+}{\tau^+ + \tau^-}u_h^+ + \frac{\tau^-}{\tau^+ + \tau^-}u_h^- + \frac{1}{\tau^+ + \tau^-}[\![\mathbf{q}_h]\!] \ . \qquad (3.26)$$

Inserting λ_h into $\hat{\mathbf{q}}_h = \mathbf{q}_h^+ - \tau^+(u_h^+ - \lambda)\mathbf{n}^+$ we get

$$\hat{\mathbf{q}}_h = \frac{\tau^-}{\tau^+ + \tau^-}\mathbf{q}_h^+ + \frac{\tau^+}{\tau^+ + \tau^-}\mathbf{q}_h^- + \frac{\tau^+\tau^-}{\tau^+ + \tau^-}[\![u_h]\!] \ . \qquad (3.27)$$

For $\tau^+ = \tau^-$ we have $\lambda_h = \{u_h\} + 1/(2\tau)[\![\mathbf{q}_h]\!]$ and $\hat{\mathbf{q}}_h = \{\mathbf{q}_h\} + \tau/2[\![u_h]\!]$. Thus the hybridized local Discontinuous Galerkin methods with $\tau^+ = \tau^-$ are equivalent to the Discontinuous Galerkin methods of [21] with $\mathbf{C}_{12} := 0$, $C_{11} := -\frac{1}{2}\tau$ and $C_{22} := -\frac{1}{2\tau}$. This means that the convergence results of [21] (for diffusion problems) transfer directly to their hybridized versions with $\tau^+ = \tau^-$.

For completeness of the presentation, we state the following theorem that has been proved in [21] for diffusion problems.

Lemma 3.3.3

Let the data of the partial differential equation (2.2) *be sufficiently regular such that* $u \in H^{k+2}(\Omega)$ *and* $\mathbf{q} \in H^{k+2}(\Omega)^2$. *Then, for the hybrid DG methods with local spaces* (3.24b) *and the numerical flux defined by* (3.25) *the following a priori estimate holds true.*

$$\|u - u_h\|_0 + h^{1/2}\|\mathbf{q} - \mathbf{q}_h\|_0 \leq Ch^{k+1}|u|_{k+2} \qquad \text{for } \tau = O(1) ,$$

$$\|u - u_h\|_0 + h \quad \|\mathbf{q} - \mathbf{q}_h\|_0 \leq Ch^{k+1}|u|_{k+2} \qquad \text{for } \tau = O(h^{-1}) .$$

We note that for the classical local DG method $\mathbf{q}_h$ is eliminated from the global system such that the global system can be formulated solely in terms of u_h. See for example [6]. Therefore, the numerical trace $\hat{u}_h$ has to be independent of the fluxes $\mathbf{q}_h$. However, for the hybridized local DG method we just have seen that $\hat{u}_h = \lambda_h$ depends on $[\![\mathbf{q}_h]\!]$ for $\tau < \infty$. This means that the classical local DG method cannot coincide with its hybridized version defined by equations (3.25) and (3.24).

As we have indicated in the motivation for the definition of the numerical fluxes at the beginning of the section these hybridized local Discontinuous Galerkin methods are related to the Discontinuous Galerkin methods analyzed in [21]. However, they are not just special cases of those methods that are suitable for hybridization. There is only one penalization parameter τ left to be set. But this parameter defined on the faces $E \in \mathcal{E}$ may differ from one side of the face to the other on interior faces. Furthermore, the polynomial degrees of the local spaces are not bound to be equal.

3.3.2 Interior Penalty DG method

Let us attempt to hybridize the Interior Penalty Discontinuous Galerkin method. Here, the procedure is not so clear as it was for the hybridization of the Local DG methods. The original idea behind IPDG methods for second order elliptic equations was to use discontinuous trial and test functions instead of continuous ones while approximating continuity via penalization of the jumps in the solution across interior faces. Originally, the notion of numerical fluxes has not been used for this method in contrast to the Local DG methods. Within the general framework for hybridization, however, we need to specify numerical fluxes $\hat{\mathbf{Q}}_m$ and $\hat{\mathbf{Q}}_f$.

In the review paper [6], a formerly not known connection between the numerical flux based DG methods and the bilinear form based DG methods has been pointed out. In this context, a numerical flux has been assigned to the Interior Penalty DG method in order to view the method in the context of flux-based DG methods. These fluxes, however, are not completely suitable for hybridization as we will see later.

In order to find suitable numerical fluxes $\hat{\mathbf{Q}}_m$ and $\hat{\mathbf{Q}}_f$, we start from the general comprehensive form of hybridized methods (3.8). We fix the discrete spaces to be

$$\mathbf{V}(T) := \mathbb{P}_k(T)^2 , \quad W(T) := \mathbb{P}_k(T) , \quad k \geq 1 . \tag{3.28}$$

Suppose $\hat{\mathbf{q}}_h|_{\mathcal{E}^0} \in M_h^0$. Then the third equation of (3.8) ensures that $\hat{\mathbf{q}}_h$ is single valued on $\mathcal{E}$. Using this fact in the second equation of (3.8), applying the integration by parts formula in the first equation of (3.8), and using relation (2.19), the system (3.8) reads

$$(a^{-1}\mathbf{q}, \mathbf{v}) + (\nabla u, \mathbf{v})_T + \langle \lambda - \{u\}, [\![\mathbf{v}]\!]\rangle_{\mathcal{E}^0} - \langle [\![u]\!], \{\mathbf{v}\}\rangle_{\mathcal{E}^0} = \langle -g_h + u, \mathbf{v}\cdot\mathbf{n}\rangle_{\partial\Omega} \quad \forall \mathbf{v} \in \mathbf{V}_h ,$$

$$-(\mathbf{q}, \nabla w)_T + (du, w) + \langle [\![w]\!], \hat{\mathbf{q}}_h\rangle_{\mathcal{E}^0} = (f, w) - \langle w, \hat{\mathbf{q}}_h\cdot\mathbf{n}\rangle_{\partial\Omega} \quad \forall w \in W_h .$$

Choosing $\mathbf{v} = a\nabla w$ in the first equation and adding the two equations above we arrive at

$$a_{HDG}(u, w) = b_{HDG}(w) \quad \forall w \in W_h,$$

where

$$b_{HDG}(w) := (f, w) - \langle g_h, a\nabla w\cdot\mathbf{n}\rangle_{\partial\Omega}$$

and

$$a_{HDG}(u, w) := (a\nabla u, \nabla w)_T + (du, w)$$
$$+ \langle \lambda - \{u\}, [\![a\nabla w]\!]\rangle_{\mathcal{E}^0} - \langle [\![u]\!], \{a\nabla w\}\rangle_{\mathcal{E}^0} + \langle \hat{\mathbf{q}}_h, [\![w]\!]\rangle_{\mathcal{E}^0} + \partial\Omega\text{-terms.} \tag{3.30}$$

Comparing this form of the hybridized DG method with the primal form of the standard IPDG method for symmetric a with some penalty parameter κ

$$a_{IP}(u, w) := (a\nabla u, \nabla w)_T + (du, w)$$
$$- \langle [\![u]\!], \{a\nabla w\}\rangle_{\mathcal{E}^0} + \langle \kappa[\![u]\!] - \{a\nabla u\}, [\![w]\!]\rangle_{\mathcal{E}^0} + \partial\Omega\text{-terms,} \tag{3.31}$$

we see that we only can recover the standard IPDG method, if we choose the numerical flux $\hat{\mathbf{q}}_h = \kappa[\![u]\!] - \{a\nabla u\}$. The numerical trace λ of the resulting H-DG method turns out to be $\{u\}$. For the local solvers, however, the numerical flux $\hat{\mathbf{q}}_h$ has to be formulated in terms of functions and data from one element T. If $\lambda_h = \{u\}$ this, however, is not possible. Hence the standard IPDG method is not hybridizable within our framework.

However, we can compare (3.30) with the stabilized symmetric IPDG method introduced in [32] and rewritten singly in terms of the primal variable u in [29]. The relevant bilinear form of this method reads

$$a_{ss}(u, w) := (a\nabla u, \nabla w)_{\mathcal{T}} + (du, w)$$
$$- \langle \frac{1}{4\kappa}[\![a\nabla u]\!], [\![a\nabla w]\!]\rangle_{\mathcal{E}^0} - \langle [\![u]\!], \{a\nabla w\}\rangle_{\mathcal{E}^0} + \langle \kappa[\![u]\!] - \{a\nabla u\}, [\![w]\!]\rangle_{\mathcal{E}^0} + \partial\Omega\text{-terms.} \quad (3.32)$$

We see that this form corresponds to (3.30), if $\hat{\mathbf{q}}_h := \kappa[\![u]\!] - \{a\nabla u\}$ and $\lambda = \{u\} - \frac{1}{4\kappa}[\![a\nabla u]\!]$. Using these representations for $\hat{\mathbf{q}}_h$ and λ, we can perform the same steps as for the derivation of the numerical fluxes for the hybridized LDG method to define the numerical flux according to

$$\hat{\mathbf{q}}_h^+ := -a\nabla u^+ + \tau(u^+ - \lambda)\cdot\mathbf{n}^+ \ ,$$
$$\hat{\mathbf{q}}_h^- := -a\nabla u^- + \tau(u^- - \lambda)\cdot\mathbf{n}^- \ .$$

Therefore, we only have to use $\beta := 0$, $\tau := 2\kappa$ and substitute $\mathbf{q}$ by $-a\nabla u$ in the derivation of the strictly local numerical fluxes of the LDG methods.

After motivating the form of the numerical flux, we now define the numerical flux of the H-IPDG method on any element $T \in \mathcal{T}$ with outward pointing normals $\mathbf{n}$ in terms of local solvers $\mathbf{Q}$ and $\mathbf{U}$ by means of

$$\hat{\mathbf{Q}}m := -a\nabla\mathbf{U}m + \tau(\mathbf{U}m - m)\mathbf{n} \ , \quad \hat{\mathbf{Q}}f := -a\nabla\mathbf{U}f + \tau(\mathbf{U}f)\mathbf{n} \ ,$$
$$M_h = \mathbb{P}_k(\mathcal{E}) := \prod_{E\in\mathcal{E}} \mathbb{P}_k(E) \ .$$

As for the LDG methods we allow the penalty parameter τ to be double-valued on each interior face E. This is a mild generalization of the stabilized IPDG method of [29]. As local spaces we choose

$$\mathbf{V}(T) := \mathbb{P}_k(T)^2 \ , \quad W(T) := \mathbb{P}_k(T) \ , \quad k \in \mathbb{N} \ , \tau > 0 \ . \quad (3.34)$$

In order to check whether the so defined method fits into our framework of hybridization, we start from the general form of local solver (3.5) and use the integration by parts formula in the first equation such that we have

$$(a^{-1}\mathbf{q}, \mathbf{v})_T + (\nabla u, \mathbf{v})_T + \langle m - u, \mathbf{v} \cdot \mathbf{n}\rangle_{\partial T} = 0 \qquad \forall \mathbf{v} \in \mathbf{V}(T) \ ,$$
$$-(\mathbf{q}, \nabla w)_T + (du, w)_T + \langle w, \hat{\mathbf{q}}_m \cdot \mathbf{n}\rangle_{\partial T} = 0 \qquad \forall w \in W(T) \ .$$

To get rid of $\mathbf{q}$ we set $\mathbf{v} = a\nabla u$, $w = u$ and add the two equations, which yields

$$(a\nabla u, \nabla u)_T + (du, u)_T + \langle m - u, a\nabla u \cdot \mathbf{n}\rangle_{\partial T} + \langle u, \hat{\mathbf{q}}_m \cdot \mathbf{n}\rangle_{\partial T} = 0 \ . \qquad (3.35)$$

This form can be used to prove the following lemma.

Lemma 3.3.4

With the local spaces defined by (3.34) and the numerical flux defined by (3.33) the local solvers (3.5) and (3.6) are well defined for $\tau > \gamma(k+1)^j h_T^{-1}$, $j \geq 1$, where k denotes the polynomial degree of the local space $W(T)$ and γ depends on the minimal angle of T.

Proof. (3.5) and (3.6) coincide for $f = 0$ and $m = 0$. Hence it suffices to show, that the homogeneous system (3.5) admits the unique solution $(\mathbf{q}, u) = 0$. We insert the definition of the H-IPDG flux $\hat{\mathbf{q}}_m := -a\nabla u + \tau(u - m) \cdot \mathbf{n}$ and set $m = 0$ in (3.35) to get

$$0 = (a\nabla u, \nabla u)_T + (du, u)_T - 2\langle u, a\nabla u \cdot \mathbf{n}\rangle_{\partial T} + \langle \tau u, u\rangle_{\partial T}$$
$$\geq (1 - \frac{(k+1)C}{\kappa})(a\nabla u, \nabla u)_T + (du, u) + (\tau - \frac{\kappa}{4h_T})\langle u, u\rangle_{\partial T} \ ,$$

where we have used Corollary 3.1.2 which holds for any $\kappa > 0$ to estimate the right hand-side of the equation from below. Now, if we set $\kappa > (k+1)C$ and $\tau > \frac{\kappa}{4h_T}$, the right-hand side of the inequality is nonnegative which implies $u = 0$. Using this information in the first equation of (3.5), we get $\mathbf{q} = 0$. $\qquad \square$

Lemma 3.3.5

Assumptions 3.1.5 and 3.1.6 are satisfied as well.

Proof. To show the local semi-definiteness of the local solver expressed in (3.11) we simply add $-\langle m, \hat{\mathbf{Q}}m \cdot \mathbf{n}\rangle_{\partial T}$ to (3.35) and again insert the actual form of the H-IPDG flux to obtain

$$-\langle m, \hat{\mathbf{Q}}m \cdot \mathbf{n}\rangle_{\partial T} = (a\nabla u, \nabla u)_T + (du, u)_T + \langle u - m, -2a\nabla u \cdot \mathbf{n} + \tau(u - m)\rangle_{\partial T}$$
$$\geq (1 - \frac{(k+1)C}{\kappa})(a\nabla u, \nabla u)_T + (du, u) + (\tau - \frac{\kappa}{4h_T})\langle u - m, u - m\rangle_{\partial T} \ ,$$

where we have used Corollary 3.1.2 again.

By the same arguments as above we conclude that $-\langle m, \hat{\mathbf{Q}}m\cdot\mathbf{n}\rangle_{\partial T} \geq 0$. The remaining parts of the assumptions are trivially fulfilled, if we choose $M^\sharp(\partial T) = \prod_{E \subset \partial T} P_k(E)$ for instance. □

In view of the definition of the numerical flux operator $\hat{\mathbf{Q}}$, the global system for the H-IPDG method can be formulated independently of the local solver $\mathbf{Q}$.

Lemma 3.3.6

With the H-IPDG numerical flux operator $\hat{\mathbf{Q}}$ defined as in (3.33) the global system $a_h(\eta, \mu) = b_h(\mu)\ \forall \mu \in M_h^0$ defined in (3.9) can be formulated as follows:

$$a_h(\eta,\mu) = (a\nabla\mathbf{U}\eta, \nabla\mathbf{U}\mu)_{\mathcal{T}} + (d\mathbf{U}\eta, \mathbf{U}\mu)_{\mathcal{T}} \tag{3.36}$$
$$+ \langle 1, [\![(\eta - \mathbf{U}\eta)a\nabla\mathbf{U}\mu + (\mu - \mathbf{U}\mu)a\nabla\mathbf{U}\eta]\!]\rangle_{\mathcal{E}^0}$$
$$+ \langle 1, [\![(\mathbf{U}\mu - \mu)\tau(\mathbf{U}\eta - \eta)\mathbf{n}]\!]\rangle_{\mathcal{E}^0}$$

and

$$b_h(\mu) = \langle g_h, [\![-a\nabla\mathbf{U}\mu + \tau(\mathbf{U}\mu - \mu)]\!]\rangle_{\partial\Omega} + (f, \mathbf{U}\mu)_{\mathcal{T}}. \tag{3.37}$$

To prove this lemma we use the following relations that are true for any hybridized method with local solvers (3.5) and (3.6). For any $m, \mu \in M_h,\ f \in L_2(\Omega), \mathbf{v} \in V_h$ and $w \in W_h$ the following identities hold

$$(a^{-1}\mathbf{Q}m + \nabla\mathbf{U}m, \mathbf{v})_{\mathcal{T}} = +\langle 1, [\![(\mathbf{U}m - m)\mathbf{v}]\!]\rangle_{\mathcal{E}}, \tag{3.38}$$
$$(\nabla\cdot\mathbf{Q}m + d\mathbf{U}m, w)_{\mathcal{T}} = -\langle 1, [\![w(\hat{\mathbf{Q}}m - \mathbf{Q}m)]\!]\rangle_{\mathcal{E}}, \tag{3.39}$$
$$(a^{-1}\mathbf{Q}f + \nabla\mathbf{U}f, \mathbf{v})_{\mathcal{T}} = +\langle 1, [\![(\mathbf{U}f)\mathbf{v}]\!]\rangle_{\mathcal{E}}, \tag{3.40}$$
$$(\nabla\cdot\mathbf{Q}f + \nabla\mathbf{U}f, w)_{\mathcal{T}} = -\langle 1, [\![w(\hat{\mathbf{Q}}f)]\!]\rangle_{\mathcal{E}}. \tag{3.41}$$

This can be seen easily by summing up the local solvers of all the elements of the triangulation and using integration by parts where necessary.

Proof. Due to (3.38) with $m := \mu$ and $\mathbf{v} := \mathbf{Q}\eta$, the bilinear form $a_h(.,.)$ of (3.9) can be formulated independently of $\mathbf{Q}\eta$. Using the same relation with $m := \eta$ and $\mathbf{v} := -\nabla\mathbf{U}\mu$, bilinear form $a_h(.,.)$ admits the desired form. Inserting $\hat{\mathbf{Q}}\mu = -a\nabla\mathbf{U}\mu + \tau(\mathbf{U}\mu - \mu)\mathbf{n}$ and $\hat{\mathbf{Q}}f = -a\nabla\mathbf{U}f + \tau\mathbf{U}f\mathbf{n}$ into the general form of $b_h(.)$ and using the fact that we can extend

g_h by zero to M_h we arrive at

$$b_h(\mu) = \langle g_h, [\![\hat{\mathbf{Q}}\mu]\!]\rangle_{\partial\Omega} + (f, \mathbf{U}\mu)_{\mathcal{T}}$$
$$+ \langle 1, [\![\mathbf{U}f(-a\nabla\mathbf{U}\mu - \mathbf{Q}\mu)]\!] - [\![(\mathbf{U}\mu - \mu)(-a\nabla\mathbf{U}f - \mathbf{Q}f)]\!]\rangle_{\mathcal{E}} \quad (3.42)$$

after cancellation of the terms $\langle 1, [\![(\mathbf{U}\mu - \mu)\mathbf{n}\tau\mathbf{U}f]\!]\rangle_{\mathcal{E}}$. Using (3.40) with $\mathbf{v} := \mathbf{Q}\mu$ as well as with $\mathbf{v} := a\nabla\mathbf{U}\mu$ in the first jump term and (3.38) with $\mathbf{v} := \mathbf{Q}f$ as well as with $\mathbf{v} := a\nabla\mathbf{U}f$ in the second jump term in the second line of $b_h(\mu)$ we note that these two jump terms cancel each other. So we arrive at the desired result. $\qquad\square$

As for the H-LDG method we get the following lemma about the precise form of the numerical fluxes.

Lemma 3.3.7
The numerical fluxes $\hat{\mathbf{q}}_h$ and $\lambda_h = \hat{u}_h$ of the H-IPDG method can be written in the form

$$\hat{u}_h = \frac{\tau^+}{\tau^+ + \tau^-}u_h^+ + \frac{\tau^-}{\tau^+ + \tau^-}u_h^- - \frac{1}{\tau^+ + \tau^-}[\![a\nabla u_h]\!] , \quad (3.43)$$

$$\hat{\mathbf{q}}_h = -\frac{\tau^-}{\tau^+ + \tau^-}a^+\nabla u_h^+ - \frac{\tau^+}{\tau^+ + \tau^-}a^-\nabla u_h^- + \frac{\tau^+\tau^-}{\tau^+ + \tau^-}[\![u_h]\!] . \quad (3.44)$$

These identities can be proved in complete analogy to the proof of the corresponding lemma for the H-LDG methods.

Of course, the special choice of penalty parameters being single valued on interior faces is included in the H-IPDG methods. Therefore we recover the symmetric stabilized IPDG method of [29] by setting $\tau^+ = \tau^- = 2\kappa$. The standard a priori error estimates for the IPDG method (3.31) apply to the H-IPDG method which is equivalent to the stabilized IPDG method of [29]. This has been noted for example in [51]. If we look at the bilinear form a_{ss} in (3.32), we notice that on one hand the presence of the gradient in the additional penalization term including the jumps in the gradient may cause a loss of accuracy of the method. This loss, however, is compensated by the penalization parameter in front of this term which is $1/(4\kappa) \approx h_T$. We omit a detailed proof of this result. So we have the following theorem (see for instance [51], Thm 2.13).

Theorem 3.3.8
Let the data of the underlying partial differential equation (2.2) be such that its solution u belongs to $H^{k+1}(\Omega)$ with $k > 1/2$. Assume also that the penalty parameter $\tau_T = O(h_T^{-1})$ is large enough. Then, there is a constant C independent of h such that the following a

priori error estimates hold:

$$\|u - u_h\|_0 + \|u - u_h\|_{1,\mathcal{T}} \leq Ch^k \|u\|_{k+1} ,$$
$$\|u - u_h\|_0 \quad \leq Ch^{k+1} \|u\|_{k+1} .$$

3.4 Hybridized continuous Galerkin method

In this section, we show that even the classical H^1-conforming finite element method can be recovered within the framework of hybridization in spite of seemingly being not well suited for hybridization.

We start from the definition of the numerical flux $\hat{\mathbf{Q}}m$ of the H-LDG method

$$\hat{\mathbf{Q}}m := \mathbf{Q}m + \tau(\mathbf{U}m - m)\mathbf{n}$$

on an arbitrary element $T \in \mathcal{T}$. If we formally pass the penalty parameter τ to infinity, which is theoretically not excluded in our definition of the H-LDG method, we see that in order to get a finite value for $\hat{\mathbf{Q}}m$, the function m has to be equal to $\mathbf{U}m \in \mathbb{P}_k(T)$ in the limit. Therefore, m has to be continuous on ∂T. Since it also is single valued on each face, we see that m is continuous on $\mathcal{E}$. Gluing together all solutions of the local solvers, we get a continuous hence H^1-conforming function $u \in H^1(\Omega)$. However, the value of $\hat{\mathbf{Q}}m$ can not be predicted, since we do not know the value of the formal product of 0 and infinity.

This argument motivates the following definition of the local problems of the Hybridized Continuous Galerkin method introduced in [25]. M_h should consist of continuous functions while the identity $\mathbf{U}m|_{\partial T} = m$ has to be satisfied strongly. Since the numerical flux $\hat{\mathbf{Q}}$ is an unknown, we have to specify a function space $\mathbf{T}(\partial T)$ for its approximation. Thus for $k \geq 1$ and any $T \in \mathcal{T}$ the local spaces are defined by

$$\mathbf{V}(T) := \mathbb{P}_{k-1}(T)^2 , \quad W(T) := \mathbb{P}_k(T) \text{ and} \qquad (3.45)$$
$$\mathbf{T}(\partial T) := \{\mathbf{n}_T w|_{\partial T} \ : \ w \in W(T)\} , \quad M_h = M_{h,k}^c := \prod_{E \in \mathcal{E}} \mathbb{P}_k(E) \cap C^0(\mathcal{E}) . \qquad (3.46)$$

The local solver $(\mathbf{Q}(\cdot), \mathbf{U}(\cdot), \hat{\mathbf{Q}}(\cdot)) : M(\partial T) \to \mathbf{V}(T) \times W(T) \times \mathbf{T}(\partial T); \ m \mapsto (\mathbf{q}_m, u_m, \hat{\mathbf{q}}_m)$ is defined by the unique solution of

$$(a^{-1}\mathbf{q}_m, \mathbf{v})_T - (u_m, \nabla \cdot \mathbf{v})_T = -\langle m, \mathbf{v} \cdot \mathbf{n}\rangle_{\partial T} \qquad \forall \mathbf{v} \in \mathbf{V}(T), \qquad (3.47a)$$
$$-(\mathbf{q}_m, \nabla w)_T + (du_m, w)_T + \langle w, \hat{\mathbf{q}}_m \cdot \mathbf{n}\rangle_{\partial T} = 0 \qquad \forall w \in W(T), \qquad (3.47b)$$
$$u_m = m \qquad \text{on } \partial T , \qquad (3.47c)$$

while the corresponding solver $(\mathbf{Q}(\cdot), \mathbf{U}(\cdot), \hat{\mathbf{Q}}(\cdot))$: $L^2(T) \to \mathbf{V}(T) \times W(T) \times \mathbf{T}(\partial T)$; $f \mapsto (\mathbf{q}_f, u_f, \hat{\mathbf{q}}_f)$ is defined by the unique solution of

$$(a^{-1}\mathbf{q}_f, \mathbf{v})_T - (u_f, \nabla\cdot\mathbf{v})_T = 0 \qquad\qquad \forall \mathbf{v} \in \mathbf{V}(T), \qquad (3.48a)$$

$$-(\mathbf{q}_f, \nabla w)_T + (du_f, w)_T + \langle w, \hat{\mathbf{q}}_f\cdot\mathbf{n}\rangle_{\partial T} = (f, w) \qquad \forall w \in W(T), \qquad (3.48b)$$

$$u_f = 0 \qquad\qquad \text{on } \partial T . \qquad (3.48c)$$

In comparison to the form of (3.5) and (3.6) the just defined local solvers have an additional equation that has to be fulfilled on each element $T \in \mathcal{T}$. Lemma 3.1.4, however, still applies, since its proof only uses the first two equations of (3.47) and (3.48). These two equations admit the same form as the corresponding equations of the general local solvers (3.5) and (3.6).

Lemma 3.4.1

With the function spaces of the H-CG method, the local solvers of the H-CG method are well defined. Furthermore, Assumptions 3.1.5 and 3.1.6 are satisfied with $M^\sharp(\partial T) := L^2(\partial T)$ for instance.

A simple proof for this lemma using the same techniques as for the corresponding proofs of H-RT or H-IPDG methods can be found in section 3.3 of [25].

Using integration by parts in (3.47a) and the interior trace condition (3.47c) together with the fact that $\nabla \mathbf{U}m$, $\mathbf{Q}m \in \mathbf{V}(T)$, we see that $\mathbf{Q}m = -a\nabla \mathbf{U}m$ whenever a is element-wise constant. Obviously, the same holds true for $\mathbf{Q}f = -a\nabla \mathbf{U}f$. After applying (3.38) and (3.40) in the general formulation for $b_h(\mu)$ in (3.9), we use these identities to get rid of the last four jump terms of $b_h(\mu)$. Hence, for the H-CG method the global system admits the form

$$(a\nabla \mathbf{U}\eta, \mathbf{U}\mu) + (d\mathbf{U}\eta, \mathbf{U}\mu) = \langle g_h, [\![\hat{\mathbf{Q}}\mu]\!]\rangle_\mathcal{E} + (f, \mathbf{U}\mu) \qquad \forall \mu \in M^c_{h,k} . \qquad (3.49)$$

In order to obtain the relationship between the H-CG method and the standard CG method, we insert $\mathbf{q}_h = -a(\nabla \mathbf{U}m + \nabla \mathbf{U}f + \nabla \mathbf{U}g_h)$ into the second equation of the general form (3.8). Since the trace of any function $w \in W_h$ is in $M^c_{h,k}$, we can also use the third equation of (3.8) in the second one to arrive at

$$(a\nabla u_h, \nabla w) + (du_h, w) = (f, w) \qquad \forall w \in \{w \in W_h : w|_{\partial\Omega} = g_h\} \cap C^0(\Omega). \qquad (3.50)$$

Thus the H-CG method is equivalent to the standard CG method for element-wise constant a. In particular, the a priori error estimates of the CG method apply which we state here for completeness.

Lemma 3.4.2

Let the data of the partial differential equation (2.2) be sufficiently regular such that $u \in H^{k+2}(\Omega)$. Then, for the CG methods with local spaces $\{w \in W_h : w|_{\partial\Omega} = g_h\} \cap C^0(\Omega)$ the following a priori estimates hold true.

$$\|u - u_h\|_1 \leq Ch^k |u|_{k+1} ,$$
$$\|u - u_h\|_0 \leq Ch^{k+1} |u|_{k+1} .$$

The proofs are standard and can be found in any book on finite element methods, e.g. [48].

3.5 Hybridized nonconforming method

As a last example, we hybridize H^1-nonconforming methods like the method introduced by Crouzeix and Raviart [27] in the context of the stationary Stokes equation. In comparison to the hybridization of the H^1-conforming methods, the situation is a bit easier here, since the approximate solution u_h is not required to be continuous across interior faces. We only have a weak continuity constraint. So we will be able to choose strictly local spaces like for the hybridization of mixed methods and the DG methods. This turns out to be easier concerning implementation.

For any odd $k \geq 1$ the local spaces are defined by

$$\mathbf{V}(T) := \mathbb{P}_{k-1}(T)^2 , \quad W(T) := \mathbb{P}_k(T) , \quad (3.51)$$

$$\mathbf{T}(\partial T) := \{\mathbf{n}_T p|_{\partial T} : p \in \mathbb{P}_{k-1}(E) \forall E \subset \partial T\} , \quad M_h = M_{h,k-1} := \prod_{E \in \mathcal{E}} \mathbb{P}_{k-1}(E) . \quad (3.52)$$

The local solver $(\mathbf{Q}(\cdot), \mathbf{U}(\cdot), \hat{\mathbf{Q}}(\cdot)) : M(\partial T) \to \mathbf{V}(T) \times W(T) \times \mathbf{T}(\partial T); \ m \mapsto (\mathbf{q}_m, u_m, \hat{\mathbf{q}}_m)$ is characterized by

$$(a^{-1}\mathbf{q}_m, \mathbf{v})_T - (u_m, \nabla \cdot \mathbf{v})_T = -\langle m, \mathbf{v} \cdot \mathbf{n} \rangle_{\partial T} \quad \forall \mathbf{v} \in \mathbf{V}(T), \quad (3.53a)$$

$$-(\mathbf{q}_m, \nabla w)_T + (du_m, w)_T + \langle w, \hat{\mathbf{q}}_m \cdot \mathbf{n} \rangle_{\partial T} = 0 \quad \forall w \in W(T), \quad (3.53b)$$

$$\langle u_m, \mu \rangle_{\partial T} = \langle m, \mu \rangle_{\partial T} \quad \forall \mu \in M(\partial T), \quad (3.53c)$$

while the corresponding solver $(\mathbf{Q}(\cdot), \mathbf{U}(\cdot), \hat{\mathbf{Q}}(\cdot)) : L^2(T) \to \mathbf{V}(T) \times W(T) \times \mathbf{T}(\partial T);$

$f \mapsto (\mathbf{q}_f, u_f, \hat{\mathbf{q}}_f)$ is characterized by

$$(a^{-1}\mathbf{q}_f, \mathbf{v})_T - (u_f, \nabla \cdot \mathbf{v})_T = 0 \qquad \forall \mathbf{v} \in V(T), \qquad (3.54a)$$

$$-(\mathbf{q}_f, \nabla w)_T + (du_f, w)_T + \langle w, \hat{\mathbf{q}}_f \cdot \mathbf{n} \rangle_{\partial T} = (f, w) \qquad \forall w \in W(T), \qquad (3.54b)$$

$$\langle u_f, \mu \rangle_{\partial T} = 0 \qquad \forall \mu \in M(\partial T). \qquad (3.54c)$$

In [25], it has been showed that the local solvers (3.53) and (3.54) are well defined for any odd $k \geq 1$. The proof is carried out by showing that the homogeneous system (3.53) admits only the trivial solution. Therefore, suitable test functions $\mathbf{v}$, w and μ are chosen using univariate orthogonal polynomials glued together continuously along the three faces of ∂T. This strategy works only for odd k.

Assumptions 3.1.5 and 3.1.6 are satisfied as well. This can be seen easily by choosing appropriate test functions and using the properties of the local spaces (see [25]). Hence, the H-NC method is well defined.

For the H-NC method the discrete jump condition

$$\langle \mu, [\![\hat{\mathbf{q}}_h]\!] \rangle_{\mathcal{E}} = 0 \quad \forall \mu \in M_h^0$$

implies $[\![\hat{\mathbf{q}}_h]\!]_E = 0$ on all interior faces of the triangulation, since $[\![\hat{\mathbf{q}}_h]\!]_E \in \mathbb{P}_{k-1}(E) = M_h|_E$.

Again the global solver of the H-NC method can be simplified. We rewrite the jump term in the general form of a_h of (3.9) in the following way

$$\langle 1, [\![(\mathbf{U}\mu - \mu)(\hat{\mathbf{Q}}\eta - \mathbf{Q}\eta)]\!] \rangle_{\mathcal{E}} = \sum_{T \in \mathcal{T}} \langle \mathbf{U}\mu - \mu, (\hat{\mathbf{Q}}\eta - \mathbf{Q}\eta) \cdot \mathbf{n} \rangle_{\partial T} .$$

Due to (3.53c) this term vanishes, since $(\hat{\mathbf{Q}}\eta - \mathbf{Q}\eta) \cdot \mathbf{n} \in M_h(\partial T)$ on every element $T \in \mathcal{T}$. Equivalently, we see that the right hand side of the global system (3.9) can be simplified to $b_h(\mu) = \langle g_h, [\![\hat{\mathbf{Q}}\mu]\!] \rangle_{\mathcal{E}} + (f, \mathbf{U}\mu)_{\mathcal{T}}$.

In order to get a simplified equivalent formulation, we use integration by parts in (3.53a) and use (3.53c) to see that for any element-wise constant a we have $\mathbf{q}_m = -a\nabla u_m$. The same can be seen for $\mathbf{q}_f$. Adding both identities, we have $\mathbf{q}_h = -a\nabla u_h$. Inserting this into the second equation of the compact form of hybridized methods (3.8) choosing and using the just established fact that the discrete jump condition is satisfied strongly, we arrive at

$$(a\nabla u_h, \nabla w)_{\mathcal{T}} + (du_h, w)_{\mathcal{T}} = (f, w) \quad \forall w \in W_h. \qquad (3.55)$$

Summing up (3.54c) together with (3.53c) across all elements of $\mathcal{T}$, we see that u_h addi-

Method	$\mathbf{V}(T)$	$W(T)$	M_h	$\hat{\mathbf{q}}_m$	$\hat{\mathbf{q}}_f$
H-RT	$\mathbb{P}_k(T)^2 + \mathbf{x}\mathbb{P}_k(T)$	$\mathbb{P}_k(T)$	$M_{h,k}$	$\mathbf{q}_m$	$\mathbf{q}_f$
H-BDM	$\mathbb{P}_k(T)^2$	$\mathbb{P}_{k-1}(T)$	$M_{h,k}$	$\mathbf{q}_m$	$\mathbf{q}_f$
H-LDG	$\mathbb{P}_k(T)^2$	$\mathbb{P}_{k-1}(T)$	$M_{h,k}$	$\mathbf{q}_m + \tau(u_m-m)\cdot\mathbf{n}$	$\mathbf{q}_m + \tau(u_f)\cdot\mathbf{n}$
H-LDG	$\mathbb{P}_k(T)^2$	$\mathbb{P}_k(T)$	$M_{h,k}$	$\mathbf{q}_m + \tau(u_m-m)\cdot\mathbf{n}$	$\mathbf{q}_m + \tau(u_f)\cdot\mathbf{n}$
H-LDG	$\mathbb{P}_{k-1}(T)^2$	$\mathbb{P}_k(T)$	$M_{h,k}$	$\mathbf{q}_m + \tau(u_m-m)\cdot\mathbf{n}$	$\mathbf{q}_m + \tau(u_f)\cdot\mathbf{n}$
H-IP	$\mathbb{P}_k(T)^2$	$\mathbb{P}_k(T)$	$M_{h,k}$	$-a\nabla u_m + \tau(u_m-m)\cdot\mathbf{n}$	$-a\nabla u_f + \tau(u_f)\cdot\mathbf{n}$
H-NC	$\mathbb{P}_{k-1}(T)^2, *$	$\mathbb{P}_k(T)$	$M_{h,k-1}$	$\hat{\mathbf{q}}_m$	$\hat{\mathbf{q}}_f$
H-CG	$\mathbb{P}_{k-1}(T)^2$	$\mathbb{P}_k(T)$	$M_{h,k}^c$	$\hat{\mathbf{q}}_m$	$\hat{\mathbf{q}}_f$

*: k odd

Table 3.2: Spaces and numerical fluxes of the methods

tionally satisfies

$$\langle u_h^+ - u_h^-, \mu\rangle_E = 0 \quad \forall \mu \in \mathbb{P}_{k-1}(E), \ E \in \mathcal{E} \ , \tag{3.56}$$

where $u_h^- := g$ on boundary edges. This is the weak continuity requirement of the H^1-nonconforming methods. The fact that the H^1-nonconforming method admits a unique solution implies that for a being element-wise constant the H-NC method is equivalent to the NC method.

Consequently, the standard a priori error estimates hold true.

Lemma 3.5.1
Let the data of the partial differential equation (2.2) be sufficiently regular such that $u \in H^{k+2}(\Omega)$. Then, for the NC methods with local spaces $\{w \in W_h : w|_{\partial\Omega} = g_h\} \cap C^0(\Omega)$ the following a priori estimates hold true.

$$\|u - u_h\|_1 \leq Ch^k|u|_{k+1} \ ,$$
$$\|u - u_h\|_0 \leq Ch^{k+1}|u|_{k+1} \ .$$

The proof is a simple variation of the original one for the stationary Stokes equation presented in [27].

An overview of all individual methods set up in this chapter is presented in Tables 3.2 and 3.3. The former summarizes the local spaces and numerical fluxes to define the local solvers while the latter includes the reformulated simplified versions of the global systems.

Method	$a_h(\eta, \mu)$	$b_h(\mu)$
H-RT	$(c\mathbf{q}_\eta, \mathbf{q}_\mu)_\mathcal{T} + (du_\eta, u_\mu)_\mathcal{T}$	$(f, u_\mu)_\mathcal{T} + \langle g_h, \mathbf{q}_\mu \cdot \mathbf{n}\rangle_{\partial\Omega}$
H-BDM	$(c\mathbf{q}_\eta, \mathbf{q}_\mu)_\mathcal{T} + (du_\eta, u_\mu)_\mathcal{T}$	$(f, u_\mu)_\mathcal{T} + \langle g_h, \mathbf{q}_\mu \cdot \mathbf{n}\rangle_{\partial\Omega}$
H-LDG	$(c\mathbf{q}_\eta, \mathbf{q}_\mu)_\mathcal{T} + (du_\eta, u_\mu)_\mathcal{T}$ $+\langle 1, [\![(u_\mu - \mu)(\tau(u_\eta - \eta)\mathbf{n})]\!]\rangle_{\mathcal{E}_h}$	$(f, u_\mu)_\mathcal{T} + \langle g_h, \mathbf{q}_\mu \cdot \mathbf{n} + \tau u_\mu\rangle_{\partial\Omega}$
H-IP	$(a\nabla u_\mu, \nabla u_\eta)_\mathcal{T} + (du_\eta, u_\mu)_\mathcal{T}$ $+\langle 1, [\![(\eta - u_\eta)a\nabla u_\mu + (\mu - u_\mu)a\nabla u_\eta]\!]\rangle_{\mathcal{E}_h}$ $+\langle 1, [\![(u_\mu - \mu)(\tau(u_\eta - \eta)\mathbf{n})]\!]\rangle_{\mathcal{E}_h}$	$(f, u_\mu)_\mathcal{T} + \langle g_h, -a\nabla u_\mu \cdot \mathbf{n} + \tau u_\mu\rangle_{\partial\Omega}$
H-NC	$(a\nabla u_\mu, \nabla u_\eta)_\mathcal{T} + (du_\eta, u_\mu)_\mathcal{T}$	$(f, u_\mu)_\mathcal{T} + \langle g_h, \hat{\mathbf{q}}_\mu \cdot \mathbf{n}\rangle$
H-CG	$(a\nabla u_\mu, \nabla u_\eta)_\mathcal{T} + (du_\eta, u_\mu)_\mathcal{T}$	$(f, u_\mu)_\mathcal{T} + \langle g_h, [\![\hat{\mathbf{q}}_\mu \cdot \mathbf{n}]\!]\rangle$

Table 3.3: Weak formulation for the approximate trace

3.6   Computational aspects

One intention to use hybridization for mixed finite element methods for the diffusion equation [4] was to reduce the size of the global linear system that has to be solved. This advantage carries over to the hybridized discontinuous Galerkin methods, in particular for higher polynomial degree k. This already has been pointed out in [25].

For a quantitative comparison of the size of the global linear system, it is useful to know the relation between the number of edges, vertices, and elements of a conforming simplicial triangulation. Since each interior edge admits two adjacent elements, and each element is limited by three edges, we have the relation $|\mathcal{T}| \approx 2/3|\mathcal{E}|$, up to boundary effects. Using Euler's polyhedron formula, we conclude that $|\mathcal{N}| \approx 1/3|\mathcal{E}|$, where $|M|$ denotes the cardinality of the set M.

The IPDG method, as well as the LDG method – using polynomials of equal degree in the primal variable and each of the components of the dual variable – can be formulated singly in terms of the primal variable u_h, using polynomials of degree at most k on each element T. Since $dim(\mathbb{P}_k(T)) = 1/2(k+1)(k+2)$, the size of the global linear system is equal to $dim(W_{h,\mathrm{IP}}) = dim(W_{h,\mathrm{L}}) \approx 1/3(k+1)(k+2)|\mathcal{E}|$, while for the H-LDG method, and the H-IPDG method, the size of the global linear system is equal to $dim(M_{h,\mathrm{H}\text{-}\mathrm{IP}}) = dim(M_{h,\mathrm{H}\text{-}\mathrm{L}}) \approx (k+1)|\mathcal{E}|$.

For the H-NC method the situation is a bit more delicate. While the space of test functions for the standard primal nonconforming Finite Element method consist of polynomials of degree at most k on each $T \in \mathcal{T}$, the continuity requirement at k points on each edge $E \in \mathcal{E}^0$ reduces the size of the global linear system of the non-hybridized method to $dim(W_{h,\mathrm{NC}}) \approx 1/3(k+1)(k+2)|\mathcal{E}| - k|\mathcal{E}|$. The global linear system of the H-NC method, however, consists of $dim(M_{h,\mathrm{H}\text{-}\mathrm{NC}}) = k|\mathcal{E}|$ equations.

For the standard continuous Galerkin method the linear system consists of $dim(W_{h,\mathrm{CG}}) = |\mathcal{N}| + (k-1)|\mathcal{E}| + 1/2(k-2)(k-1)|\mathcal{T}| \approx 1/3(k^2+6)|\mathcal{E}|$ equations, where $|\mathcal{N}|$, $(k-1)|\mathcal{E}|$ and $1/2(k-2)(k-1)|\mathcal{T}|$ correspond to the nodes on the vertices, edges and to the nodes inside the elements of the triangulation, respectively. Thus, the H-CG method amounts to the solution of an equation of dimension $dim(M_{h,\mathrm{H-CG}}) \approx (k-2/3)|\mathcal{E}|$.

For the mixed methods, a discrete variational formulation can not be constructed only in terms of the primal variable. For this reason, the global linear system of the hybrid mixed methods is always smaller than the system of the original method.

k	r_{DG}	r_{NC}	r_{CG}
1	1	1	1
2	1.3	-	1
3	1.7	1.2	1.3
$\vdots$	$\vdots$	$\vdots$	$\vdots$
7	3	2.1	2.6
8	3.3	-	2.9
9	3.7	2.8	3.2

Table 3.4: approximate ratio $r_{\mathrm{meth}} := dim(W_{h,\mathrm{meth}})/dim(M_{h,\mathrm{H-meth}})$

In table 3.4, we list the ratio $r_{\mathrm{meth}} := dim(W_{h,\mathrm{meth}})/dim(M_{h,\mathrm{H-meth}})$ between the dimensions of the linear systems of a method and its hybridized version for the first polynomial degrees up to boundary effects. We observe that – in terms of globally coupled degrees of freedom – a method is never preferable in comparison to its hybridized counterpart. For larger k the ratio r_{meth} is clearly in favor of the hybrid version for any of the methods under consideration.

Chapter 4

A posteriori error analysis

4.1 A posteriori error estimators

An a posteriori error estimator that inherits information about the local distribution of the estimated error is an essential part of adaptive finite element methods. In the following we prove reliability and efficiency of the residual based a posteriori error estimator η, defined by

$$\eta^2 := \sum_{T \in \mathcal{T}} \left(\eta_{T,1}^2 + \eta_{T,2}^2 \right) + \sum_{E \in \mathcal{E}^0} \left(\eta_{E,1}^2 + \eta_{E,2}^2 \right) + \sum_{E \in \mathcal{E}^\partial} \eta_{E,3}^2 , \tag{4.1}$$

where

$$\eta_{T,1} := \| a^{-1} \mathbf{q}_h + \nabla u_h \|_{0,T}, \qquad\qquad T \in \mathcal{T} , \tag{4.2a}$$

$$\eta_{T,2} := \frac{h_T}{k+1} \| f - \nabla \cdot \mathbf{q}_h - d u_h \|_{0,T}, \qquad\qquad T \in \mathcal{T} , \tag{4.2b}$$

$$\eta_{E,1} := \left(\frac{h_E}{k+1} \right)^{1/2} \| [\![\mathbf{q}_h]\!] \|_{0,E}, \qquad\qquad E \in \mathcal{E}_h^0 , \tag{4.2c}$$

$$\eta_{E,2} := \frac{k+1}{h_E^{1/2}} \| [\![u_h]\!] \|_{0,E}, \qquad\qquad E \in \mathcal{E}_h^0 , \tag{4.2d}$$

$$\eta_{E,3} := \frac{k+1}{h_E^{1/2}} \| g_D - u_h \|_{0,E}, \qquad\qquad E \in \mathcal{E}_h^\partial . \tag{4.2e}$$

This estimator is applicable to all kinds of hybridized methods introduced in Chapter 3. Since some of the methods are equivalent to C^0- or $H(div)$-conforming methods, we point out that corresponding estimator terms cancel out a priori.

Theorem 4.1.1

Let $(\mathbf{q}_h, u_h, \lambda_h) \in \mathbf{V}_h \times W_h \times M_h$ be the solution of any of the hybridized methods of Table

3.2. In the case of the H-LDG method let $\min\{\tau^+, \tau^-\} \lesssim h_E^{-1}$. Then, the error with respect to the solution $(\mathbf{q}, u) \in \mathbf{V} \times W$ of the primal mixed formulation (2.4) can be bounded by η defined in (4.1), where $\mathbf{V} := (L^2(\Omega))^2$ and $W := H^1(\Omega)$:

$$\|\mathbf{q} - \mathbf{q}_h\|_{0,\Omega}^2 + \|u - u_h\|_{1,\mathcal{T}}^2 \lesssim \eta^2 \ . \tag{4.3}$$

The proof will be given within the following framework: In order to use a residual error concept we introduce a W-conforming approximation $\tilde{u}_h$ to the primal variable of the solution $(\mathbf{q}, u)$ of (2.2), which satisfies the Dirichlet boundary condition. In general, the computed solution u_h is nonconforming. Therefore, we introduce the consistency error ξ_i, $i = 0, 1$ according to

$$\xi_i := \min_{\tilde{v}_h \in W} \|u_h - \tilde{v}_h\|_{i,\mathcal{T}}, \ i = 0, 1. \tag{4.4}$$

Subtracting $\mathcal{L}(\mathbf{q}_h, \tilde{u}_h)$ from the operator-theoretic formulation (2.6) of the primal mixed formulation, we have

$$\mathcal{L}(\mathbf{q} - \mathbf{q}_h, u - \tilde{u}_h) = (Res_1, Res_2)^T \ ,$$

where (Res_1, Res_2) denotes the residuals $Res_1 : V \to V^*$ and $Res_2 : W_0 \to W_0^*$ of the two determining equations with respect to $(\mathbf{q}_h, \tilde{u}_h) \in \mathbf{V} \times W$, defined by

$$Res_1(\mathbf{v}) := (\nabla \tilde{u}_h + a^{-1}\mathbf{q}_h, \mathbf{v})_{0,\Omega} \ , \quad \mathbf{v} \in \mathbf{V} \tag{4.5}$$

$$Res_2(w) := (f - d\tilde{u}_h, w)_{0,\Omega} + (\mathbf{q}_h, \nabla w)_{0,\Omega} \ , \quad w \in W \ . \tag{4.6}$$

We estimate the right-hand side first in terms of the computed solution before we estimate the difference between the computed solution and the thereof computed conforming approximation $(\mathbf{q}_h, \tilde{u}_h)$.

Lemma 4.1.2
Let $(\mathbf{q}_h, u_h, \lambda_h) \in \mathbf{V}_h \times W_h \times M_h$ be the solution of the hybrid method (3.8). Choosing $\tilde{u}_h$ as the unique minimizer of (4.4), we have

$$\|Res_1\|_{\mathbf{V}^*} \lesssim (\sum_{T \in \mathcal{T}} \eta_{T,1}^2)^{1/2} + \xi_1 \ . \tag{4.7}$$

Proof. Res_1 can be estimated by

$$
\begin{aligned}
Res_1(\mathbf{v}) &= (\nabla\tilde{u}_h + a^{-1}\mathbf{q}_h, \mathbf{v})_{0,\Omega} \\
&= (\nabla u_h + a^{-1}\mathbf{q}_h, \mathbf{v})_{0,\mathcal{T}} + (\nabla(\tilde{u}_h - u_h), \mathbf{v})_{0,\Omega} \\
&\leq ((\sum_{T\in\mathcal{T}} \underbrace{\|\nabla_h u_h + a^{-1}\mathbf{q}_h\|_{0,T}^2}_{\eta_{T,1}})^{1/2} + \xi_1)\|\mathbf{v}\|_{0,\mathcal{T}}, \ \mathbf{v}\in\mathbf{V} .
\end{aligned}
\tag{4.8}
$$

$\square$

Lemma 4.1.3

Let $(\mathbf{q}_h, u_h, \lambda_h) \in \mathbf{V}_h \times W_h \times M_h$ be the solution of the H-LDG method with $k_2 = 0$. Choosing $\tilde{u}_h$ as the unique minimizer of (4.4), we can bound the norm of Res_2 by

$$
\|Res_2\|_{W_0^*} \lesssim (\sum_{T\in\mathcal{T}}\eta_{T,2}^2 + \sum_{E\in\mathcal{E}^0}\eta_{E,1}^2 + \sum_{E\in\mathcal{E}^0}(h_E\tau_m\eta_{E,2})^2 + \sum_{E\in\mathcal{E}^\partial}(h_E\tau\eta_{E,3})^2)^{1/2} + \|d\|_\infty\xi_0 ,
\tag{4.9}
$$

where $\tau_m = \min\{\tau^+, \tau^-\}$ is the minimum of the penalty parameters of the elements of both sides of the interior face E. For all other hybrid methods of Table 3.2, the norm of Res_2 can be bounded by

$$
\|Res_2\|_{W_0^*} \lesssim (\sum_{T\in\mathcal{T}}\eta_{T,2}^2 + \sum_{E\in\mathcal{E}^0}\eta_{E,1}^2)^{1/2} + \|d\|_\infty\xi_0 .
\tag{4.10}
$$

Proof. First we prove estimate (4.10). From (3.8c) we know that $\hat{\mathbf{q}}_h$ is single valued on $\mathcal{E}$. Thus, equation (3.8a) can be written in the form

$$
(f - du_h, w_h)_{0,\mathcal{T}} + (\mathbf{q}_h, \nabla w_h)_{0,\mathcal{T}} - \langle\hat{\mathbf{q}}_h, [\![w_h]\!]\rangle_{\mathcal{E}} = 0 , \qquad w_h \in W_h .
$$

For all $w_h \in W_1^c \subset W_h$ the jump term vanishes and we can write the residual as

$$
\begin{aligned}
Res_2(w) &= (f - d\tilde{u}_h, w - w_h)_{0,\mathcal{T}} + (\nabla w, \mathbf{q}_h)_{0,\mathcal{T}} \\
&= (f - du_h, w - w_h)_{0,\mathcal{T}} + (\nabla(w - w_h), \mathbf{q}_h)_{0,\mathcal{T}} + (d(u_h - \tilde{u}_h), w)_{0,\mathcal{T}} \\
&\leq (f - du_h - \nabla\!\cdot\!\mathbf{q}_h, w - w_h)_{0,\mathcal{T}} + \|d\|_\infty\xi_0\|w\|_{0,\mathcal{T}} + \langle[\![\mathbf{q}_h]\!], w - w_h\rangle_{\mathcal{E}^0} \\
&\lesssim (\sum_{T\in\mathcal{T}}\underbrace{\frac{h_T}{k+1}\|f - du_h - \nabla\!\cdot\!\mathbf{q}_h\|_{0,T}}_{\eta_{T,2}} + \sum_{E\in\mathcal{E}_h^0}\underbrace{(\frac{h_E}{k+1})^{1/2}\|[\![\mathbf{q}_h]\!]\|_{0,E}}_{\eta_{E,1}} + \|d\|_\infty\xi_0)\|w\|_{1,\mathcal{T}} ,
\end{aligned}
$$

where we have used the k dependent version of the standard finite element interpolation properties (see, e.g., [42]) of the form

$$\|w - w_h\|_{0,T} \lesssim \frac{h_T}{k+1}\|\nabla w\|_{0,\omega_T} \tag{4.11}$$

$$\|w - w_h\|_{0,E(T)} \lesssim (\frac{h_E}{k+1})^{1/2}\|w\|_{1,\omega_T} \tag{4.12}$$

$$\tag{4.13}$$

in the last step.

For lowest order mixed methods, however, this approach does not work since $W_1^c \not\subseteq W_h$. Instead, we note that both $\mathbf{q}, \mathbf{q}_h \in H(div; \Omega)$ and that $\hat{\mathbf{q}}_h = \mathbf{q}_h|_{\mathcal{E}}$. After recognizing that $[\![u_h]\!] = -[\![u - u_h]\!]$ we can use integration by parts to estimate $Res_2(w)$ in the following way:

$$\begin{aligned}
Res_2(w) &= (f - d\tilde{u}_h, w)_{0,\mathcal{T}} + (\nabla w, \mathbf{q}_h)_{0,\mathcal{T}} \\
&= (f - du_h, w - w_h)_{0,\mathcal{T}} + (\nabla(w - w_h), \mathbf{q}_h)_{0,\mathcal{T}} + \langle \hat{\mathbf{q}}_h, [\![w_h]\!] \rangle_{\mathcal{E}} + (d(u_h - \tilde{u}_h), w)_{0,\mathcal{T}} \\
&= (f - du_h - \nabla\cdot\mathbf{q}_h, w - w_h)_{0,\mathcal{T}} + (d(u_h - \tilde{u}_h), w)_{0,\mathcal{T}} \\
&\leq (\sum_{T \in \mathcal{T}} \underbrace{h_T\|f - du_h - \nabla\cdot\mathbf{q}_h\|_{0,T}}_{\eta_{T,2}} + \|d\|_\infty \xi_0)\|w\|_{1,\mathcal{T}} ,
\end{aligned}$$

where we have used the fact that for each $w \in H^1(T)$ there is an $\alpha \in P_0(T)$ such that

$$\|w - \alpha\|_{0,T} \leq Ch_T\|\nabla w\|_{0,T} , \tag{4.14a}$$

$$\|w - \alpha\|_{0,E} \leq C^* h_T\|w\|_{1,T} . \tag{4.14b}$$

With $\alpha := \int_T w dx$, (4.14a) is the Poincaré inequality, for which it has been shown in [47] that the constant C can be bounded independently of the shape of the individual element T by $C_T \leq h_T/\pi$. In this case, there are no undetermined constants present. The second estimate is a direct consequence of the first estimate and an inverse trace inequality.

The proof is least straightforward for the hybridized LDG method with $W(T) = \mathbb{P}_0(T)$. In this case, both previous approaches do not work. Starting from (3.8b) and using the single valuedness of the numerical flux, we get the following identity which is valid for all $w_h \in W_h$

$$(\mathbf{q}_h, \nabla w_h)_{0,\mathcal{T}} + (f - du_h, w_h)_{0,\mathcal{T}} - \langle \hat{\mathbf{q}}_h, [\![w_h]\!] \rangle_{\mathcal{E}} = 0 . \tag{4.15}$$

After applying the integration by parts formula once, using identity (2.19) and the explicit expression for the numerical flux $\hat{\mathbf{q}}_h$ of the hybridized LDG method (3.27), we obtain

$$0 = (f - du_h - \nabla\cdot\mathbf{q}, -w_h)_{0,\mathcal{T}} + \langle [\![\mathbf{q}_h]\!], -\{w_h\}\rangle_{\mathcal{E}} + \langle \{\mathbf{q}_h\}, -[\![w_h]\!]\rangle_{\mathcal{E}^0}$$
$$+ \langle \frac{\tau^-}{\tau^- + \tau^+}\mathbf{q}_h^+ + \frac{\tau^+}{\tau^- + \tau^+}\mathbf{q}_h^- + \frac{\tau^+\tau^-}{\tau^- + \tau^+}[\![u_h]\!], [\![w_h]\!]\rangle_{\mathcal{E}^0} + \langle \mathbf{q}_h\cdot\mathbf{n} + \tau(u_h - g_h), w_h\rangle_{\mathcal{E}^\partial} \ ,$$

which can be simplified to

$$0 = (f - du_h - \nabla\cdot\mathbf{q}, -w_h)_{0,\mathcal{T}} + \langle [\![\mathbf{q}_h]\!], -\{w_h\}\rangle_{\mathcal{E}^0}$$
$$+ \langle \frac{\tau^- - \tau^+}{\tau^- + \tau^+}\mathbf{n}^+[\![\mathbf{q}_h]\!] + \frac{\tau^+\tau^-}{\tau^- + \tau^+}[\![u_h]\!], [\![w_h]\!]\rangle_{\mathcal{E}^0} + \langle \tau(u_h - g_h), -w_h\rangle_{\mathcal{E}^\partial} \ . \quad (4.16)$$

We insert this identity into $Res_2(w)$ to arrive at

$$Res_2(w) = (f - du_h - \nabla\cdot\mathbf{q}, w - w_h)_{0,\mathcal{T}} + \langle [\![\mathbf{q}_h]\!], \{w - w_h\}\rangle_{\mathcal{E}^0} + (d(u_h - \tilde{u}_h), w)_{0,\mathcal{T}}$$
$$+ \langle \frac{\tau^- - \tau^+}{\tau^- + \tau^+}\mathbf{n}^+[\![\mathbf{q}_h]\!] + \frac{\tau^+\tau^-}{\tau^- + \tau^+}[\![u_h]\!], [\![w - w_h]\!]\rangle_{\mathcal{E}^0} + \langle \tau(u_h - g_h), w_h\rangle_{\mathcal{E}^\partial} \ ,$$

where we already used the fact that $w - \{w_h\} = \{w - w_h\}$ and $[\![w]\!] = [\![w - w_h]\!]$ on interior edges $E \in \mathcal{E}^0$. Using the Cauchy-Schwarz inequality and (4.14), this yields the following estimate:

$$Res_2(w) \lesssim \left(\|h_{\mathcal{T}}(f - du_h - \nabla\cdot\mathbf{q}_h)\|_{0,\mathcal{T}} + \xi_0 + \|h_{\mathcal{E}}^{1/2}[\![\mathbf{q}_h]\!]\|_{0,\mathcal{E}^0} \right.$$
$$\left. + \|h_{\mathcal{E}}^{1/2}\frac{\tau^+\tau^-}{\tau^+ + \tau^-}\mathbf{n}^+[\![u_h]\!]\|_{0,\mathcal{E}^0} + \|h_{\mathcal{E}}^{1/2}\tau(u_h - g_D)\|_{0,\mathcal{E}^\partial} \right)\|w\|_{1,\Omega} \ ,$$

which implies the desired result since $\frac{\tau^+\tau^-}{\tau^+ + \tau^-} \leq \min\{\tau^+, \tau^-\}$. $\qquad\square$

In order to get a computable error estimate, it is necessary to bound the actual value of ξ_0 and ξ_1.

Lemma 4.1.4

There exists an interpolation operator $\pi_{\mathcal{T}} : W_h \mapsto W_h(\mathcal{T}) \cap H_0^1(\Omega)$ such that

$$\|u_h - \pi_{\mathcal{T}}u_h\|_{0,\mathcal{T}} \lesssim \|h_{\mathcal{E}}^{1/2}[\![u_h]\!]\|_{0,\mathcal{E}} \ , \quad (4.17)$$

$$\|\nabla(u_h - \pi_{\mathcal{T}}u_h)\|_{0,\mathcal{T}} \lesssim \|\frac{k+1}{h_{\mathcal{E}}^{1/2}}[\![u_h]\!]\|_{0,\mathcal{E}} \quad (4.18)$$

where the hidden constants depend only on the local geometry of the triangulation.

Proof. A proof of (4.18) has been provided in Proposition 5.4 of [36]. It uses an explicit nodal based construction of a function $\tilde{u}_h := \pi_{\mathcal{T}}u_h \in W_{k+1}^c$. The value of $\tilde{u}_h$ at the vertices

$\mathbf{x}_V \in \mathcal{N}$ of the triangulation is defined according to

$$\tilde{u}_h(\mathbf{x}_V) := 1/|\omega_{\mathbf{x}_V}| \sum_{T \in \omega_T} u_h(\mathbf{x}_V)|_T \, ,$$

where $|\Omega_T|$ denotes the number of elements in ω_T. For the nodes located on the interior of the edges $\mathbf{x}_E \in \mathcal{E}^0 \setminus \mathcal{N}$, the value of $\tilde{u}_h$ is defined by means of

$$\tilde{u}_h(\mathbf{x}_E) := 1/2(u_h^+(\mathbf{x}_E) + u_h^-(\mathbf{x}_E)) \, , \tag{4.19}$$

while the value of $\tilde{u}_h$ at the nodes $\mathbf{x}_T \in T$ inside the elements is defined by $\tilde{u}_h(\mathbf{x}_T) := u_h(\mathbf{x}_T)$.

To estimate $\|u_h - \pi_\mathcal{T} u_h\|_{0,\mathcal{T}}$ for $k \leq 2$, we use $\tilde{u}_1$, the minimizer of $\|u_h - \pi_\mathcal{T} u_h\|_{1,\mathcal{T}}$ in W_1^c, choose a function $u_0 \in \prod_{T \in \mathcal{T}} \mathbb{P}_1(T)$ defined by $u_0|_T := \int_T (u_h - \tilde{u}_1)dx$, write

$$\|u_h - \pi_\mathcal{T} u_h\|_{0,\mathcal{T}} \leq \|u_h - \tilde{u}_1 - u_0\|_{0,\mathcal{T}} + \|\tilde{u}_1 + u_0 - \tilde{u}_h\|_{0,\mathcal{T}} \, , \tag{4.20}$$

use estimate (4.14a), the estimate for $\|u_h - \pi_\mathcal{T} u_h\|_{1,\mathcal{T}}$ and the estimate for $\|u_h - \pi_\mathcal{T} u_h\|_{0,\mathcal{T}}$ for discontinuous piecewise linear functions (see e.g.: [9] Lemma 2.2) to see that we have the estimate $\|u_h - \pi_\mathcal{T} u_h\|_{0,\mathcal{T}} \lesssim h_E^{1/2} \|[u_h]\|_{0,\mathcal{E}}$.

For $k > 2$, we make use of the regular distribution of nodes on the elements. We

Figure 4.1: Regular distribution of the basis nodes for $k + 1 = 5$ on the reference element.

have $\tilde{u}_h(\mathbf{x}_T) - u_h(\mathbf{x}_T) = 0$ on the skeleton of the simplicial submesh joining the nodes $\mathbf{x}_T$ (dashed lines in Figure 4.1). So an application of the Fundamental Theorem of Calculus on suitably chosen subdomains of $\hat{T}$ implies

$$\|u_h - \pi_\mathcal{T} u_h\|_{0,\mathcal{T}} \lesssim \frac{h_T}{k+1} \|u_h - \pi_\mathcal{T} u_h\|_{1,\mathcal{T}} \, , \ k > 2 \, . \tag{4.21}$$

Using a simple scaling argument, we see that there is a constant C, which only depends on the local geometry of the triangulation such that

$$\|u_h - \pi_\mathcal{T} u_h\|_{0,\mathcal{T}} \leq C \frac{h_T}{k+1} \|u_h - \pi_\mathcal{T} u_h\|_{1,\mathcal{T}} \, . \tag{4.22}$$

Together with (4.18) this implies (4.17). $\qquad\square$

Remark 4.1.5

For a construction of a conforming approximation $\tilde{u}_h$ satisfying Dirichlet boundary conditions $\pi_{\mathcal{T}} u_h = g_D$ on Γ, the estimates read

$$\|u_h - \pi_{\mathcal{T}} u_h\|_{0,\mathcal{T}} \lesssim \|h^{1/2}[u_h]\|_{\mathcal{E}^0} + \|h^{1/2}(u_h - g_D)\|_{\mathcal{E}^\partial} , \tag{4.23}$$

$$\|u_h - \pi_{\mathcal{T}} u_h\|_{1,\mathcal{T}} \lesssim \|\frac{k+1}{h^{1/2}}[u_h]\|_{\mathcal{E}^0} + \|\frac{k+1}{h^{1/2}}(u_h - g_D)\|_{\mathcal{E}^\partial} , \tag{4.24}$$

where the hidden constants depend only on the local geometry of the triangulation.

Proof of Theorem 4.1.1. Using $u_h = \tilde{u}_h + (u_h - \tilde{u}_h)$, the triangle inequality yields

$$\|u - u_h\|_{1,\mathcal{T}} + \|\mathbf{q} - \mathbf{q}_h\|_{0,\Omega} \leq \|u - \tilde{u}_h\|_{1,\mathcal{T}} + \|\mathbf{q} - \mathbf{q}_h\|_{0,\Omega} + \xi_1$$
$$= \|(\mathbf{q} - \mathbf{q}_h, u - \tilde{u}_h)\|_{\mathbf{V} \times W} + \xi_1 .$$

Since $u - \tilde{u}_h \in W_0$, Theorem 2.1.1 can be used, to conclude that the norm of the error can be bounded by

$$\|(\mathbf{q} - \mathbf{q}_h, u - \tilde{u}_h)\|_{\mathbf{V} \times W_0} \leq C_P \|(Res_1, Res_2)\|_{\mathbf{V}^* \times W_0^*} ,$$

where $C_P \leq 2\max\{\|a\|_\infty, \|a^{-1}\|_\infty (C_{PF}^2 + 1)\}$, and C_{PF} is the constant in the Poincaré-Friedrichs inequality. Using Lemma 4.1.2 and Lemma 4.1.3, we get

$$\|u - u_h\|_{1,\mathcal{T}} + \|\mathbf{q} - \mathbf{q}_h\|_{0,\Omega} \lesssim$$
$$\Big(\sum_{T\in\mathcal{T}} \big(\eta_{T,1}^2 + \eta_{T,2}^2\big) + \sum_{E\in\mathcal{E}^0} \big(\eta_{E,1}^2 + \tau_m^2 \eta_{E,2}^2\big) + \sum_{E\in\mathcal{E}^\partial} \tau^2 \eta_{E,3}^2 \Big)^{1/2} + \xi_1 + \|d\|_\infty \xi_0 , \tag{4.25}$$

where τ_m and τ are the parameters of Lemma 4.1.3. For all but the lowest order H-LDG method the corresponding terms vanish.

Bounding $\xi_i \leq \|u_h - \pi_{\mathcal{T}} u_h\|_{i,\mathcal{T}}$, $0 \leq i \leq 1$, we can apply Lemma 4.1.4 to complete the proof on any sufficiently fine mesh as long as $\tau_m, \tau \lesssim h_E^{-1}$ for the lowest order H-LDG method. $\qquad\square$

Remark 4.1.6

An alternative to using the bound established in Lemma 4.1.4 is to calculate the H^1-conforming function $\tilde{u}_h$ explicitly in a post-processing routine. In this case, we achieve sharper upper bounds for the consistency errors ξ_0 and ξ_1. Having an explicit $\tilde{u}_h$ at hand, a further option is to accept $\tilde{u}_h$ as a conforming numerical approximation to the primal variable of equation (2.4). In this case, only ξ_0 has to be included in the a posteriori error estimator.

We get even better estimates, if we use local post-processing to construct a better nonconforming approximation u_h^* to the primal variable before constructing a conforming approximation $\tilde{u}_h$ thereof. For hybridized mixed methods a construction using the values of λ_h has been introduced in [4]. The constructions of [53, 41, 11, 58] are based on the solution of local problems on each element $T \in \mathcal{T}$ using $\mathbf{q}_h$, u_h and f such that ∇u_h^* is close to $\mathbf{q}_h$.

For the lowest order RT mixed approximation $(\mathbf{q}_h, u_h)$ the function $u_h^*|_T \in \mathbb{P}_2(T)$ can be easily calculated, by requiring $-a\nabla u_h^*|_T = \mathbf{q}_h|_T$ and $(u_h^* - u_h, 1) = 0$. $\tilde{u}_h \in H^1(\Omega)$ is then explicitly constructed, using the strategy mentioned in the proof of Lemma 4.1.4. Accepting $\tilde{u}_h$ as the approximation for the exact solution $u \in H^1(\Omega)$, $\eta_{T,1}$ has to be replaced by $\|a^{-1}\mathbf{q}_h + \nabla\tilde{u}_h\|_{0,T}$, while $\eta_{\mathcal{E},2}$ and $\eta_{\mathcal{E},3}$ can be replaced by $\xi_0 \leq \|d(\tilde{u}_h - u_h)\|_{0,\mathcal{T}}$.

Remark 4.1.7

Using the decomposition of $a^{-1}\mathbf{q}_h \in L^2(\Omega)$ into $a^{-1}\mathbf{q}_h = \nabla\alpha + \mathrm{Curl}\beta$, where $\alpha \in H_0^1(\Omega)$, $\beta \in H^1(\Omega)$, and choosing $\tilde{u}_h := -\alpha$ it has been shown in [17] that in case of $d = 0$ for the homogeneous Dirichlet boundary value problem $\|Res_1\|_{\mathbf{V}^*}$ can be bounded by

$$\|Res_1\|_{\mathbf{V}^*} \lesssim \mu := \|h_{\mathcal{T}}\mathrm{curl}\mathbf{q}_h\|_{0,\mathcal{T}} + \|h_{\mathcal{E}}(\mathbf{q}_h^+ - \mathbf{q}_h^-) \cdot t_{\mathcal{E}}\|_{0,\mathcal{E}}, \qquad (4.26)$$

where $t_{\mathcal{E}}|_E$ denotes the tangential unit vector along $E \in \mathcal{E}$. This implies that

$$\|u - \tilde{u}_h\|_{1,\Omega} + \|\mathbf{q} - \mathbf{q}_h\|_{0,\Omega} \lesssim \sum_{T \in \mathcal{T}} \eta_{T,2}^2 + \sum_{E \in \mathcal{E}^0} \eta_{E,1}^2 + \mu.$$

To the author's knowledge, this result does not extend to the case $d \neq 0$, since the choice $\tilde{u}_h = \alpha$ does not necessarily satisfy the consistency bound $\|u_h - \alpha\|_{0,\mathcal{T}} \lesssim \|(k+1)h_{\mathcal{E}}^{1/2}[\![u_h]\!]\|_{0,\mathcal{E}^0}$.

4.2 Estimation of $\eta_{E,2}$ and $\eta_{E,3}$ for the H-DG methods

In the following, we show that for the hybridized IPDG method the jumps in the primal variable u_h can be bounded by the remaining estimator terms. Therefore we proceed in the spirit of [9]. First, we will derive a primal formulation of the H-IPDG method by rewriting equation (3.8). We will show that the corresponding bilinear form is coercive with respect to a suitable mesh dependent norm. Estimation of the difference between u_h and a conforming function $v = \pi_{\mathcal{T}}u_h \in W_h \cap H_0^1(\Omega)$ will lead us to the desired estimate.

Using integration by parts in equation (3.8a), recalling $\hat{\mathbf{q}}_h$ and $\hat{u}_h$, setting $\mathbf{v} := a\nabla w_h$, and adding equations (3.8a) and (3.8b) we get

$$
(a\nabla u_h, \nabla w)_{0,\mathcal{T}} + (du_h, w)_{0,\mathcal{T}} + \langle \hat{\mathbf{q}}_h, [\![w]\!] \rangle_{\mathcal{E}} + \langle \hat{u}_h - \{w_h\}, [\![a\nabla w]\!] \rangle_{\mathcal{E}^0}
$$
$$
- \langle [\![u_h]\!], \{a\nabla w\} \rangle_{\mathcal{E}^0} = (f, w)_{0,\mathcal{T}} + \langle u_h - g_h, a\nabla w\cdot \mathbf{n} \rangle_\Gamma \ . \quad (4.27)
$$

Using the explicit representations of the numerical traces $\hat{\mathbf{q}}_h$ and $\hat{u}_h$ (see (3.43))

$$
\hat{\mathbf{q}}_h = -\frac{\tau^-}{\tau^+ + \tau^-}a^+ \nabla u_h^+ - \frac{\tau^+}{\tau^+ + \tau^-}a^- \nabla u_h^- + \frac{\tau^+ \tau^-}{\tau^+ + \tau^-}[\![u_h]\!] \ ,
$$
$$
\hat{u}_h = \frac{\tau^+}{\tau^+ + \tau^-}u_h^+ + \frac{\tau^-}{\tau^+ + \tau^-}u_h^- - \frac{1}{\tau^+ + \tau^-}[\![a\nabla u_h]\!] \ ,
$$

on each interior edge and $\hat{\mathbf{q}}_h = -a\nabla u_h + \tau(u_h - g_h)\cdot\mathbf{n}$ on $\mathcal{E}^\partial$, we obtain the following formulation

$$
\mathcal{A}(u_h, w) = F(w) \ , \ \forall w \in W_h \quad (4.28)
$$

as the characterising equation for u_h, where

$$
\mathcal{A}(u_h, w) := (a\nabla u_h, \nabla w)_{0,\mathcal{T}} + (du_h, w)_{0,\mathcal{T}} - \langle a\nabla u_h \mathbf{n} - \tau u_h, w \rangle_{\mathcal{E}^\partial}
$$
$$
- \langle \frac{\tau^-}{\tau^+ + \tau^-}a^+ \nabla u_h^+ + \frac{\tau^+}{\tau^+ + \tau^-}a^- \nabla u_h^- - \frac{\tau^+ \tau^-}{\tau^+ + \tau^-}[\![u_h]\!], [\![w]\!] \rangle_{\mathcal{E}^0}
$$
$$
+ \langle \frac{\tau^+ - \tau^-}{2(\tau^+ + \tau^-)}\mathbf{n}^+ [\![u_h]\!] - \frac{1}{\tau^+ + \tau^-}[\![a\nabla u_h]\!], [\![a\nabla w]\!] \rangle_{\mathcal{E}^0} - \langle [\![u_h]\!], \{a\nabla w\} \rangle_{\mathcal{E}^0} \quad (4.29)
$$

and

$$
F(w) := (f, w)_{0,\mathcal{T}} + \langle u_h - g_h, a\nabla w\cdot\mathbf{n} \rangle_\Gamma + \langle \tau g_h, w \rangle_{\mathcal{E}^\partial} \ . \quad (4.30)
$$

Lemma 4.2.1 (Coercivity of $\mathcal{A}$)
Let the penalty parameter of the H-IPDG method be given by $\tau := \gamma(k + 1)^2 h_T$. Consider the mesh dependent norm $|||\cdot|||_h$ on W_h defined as

$$
|||u|||_h := \|a^{1/2}\nabla u\|_{\mathcal{T},0} + \gamma\|(k + 1)h^{1/2}[\![u]\!]\|_{\mathcal{E},0} \ , \quad (4.31)
$$

and the bilinear form $\mathcal{A}$ defined in (4.29). Then, $\mathcal{A}$ is coercive with respect to the mesh dependent norm $|||\cdot|||_h$.

Proof. Since $\frac{\tau^\pm}{\tau^+ + \tau^-} \le 1$, $\frac{\tau^+ - \tau^-}{2(\tau^+ + \tau^-)} \ge -\frac{1}{2}$ and $\frac{\tau^+ \tau^-}{\tau^+ + \tau^-} \ge \frac{\tau_m}{2}$, where $\tau_m := \min\{\tau^+, \tau^-\}$ we can write

$$
\begin{aligned}
\mathcal{A}(u_h, u_h) =\ & (a\nabla u_h, \nabla u_h)_{0,\mathcal{T}} + (du_h, u_h)_{0,\mathcal{T}} + \langle \tau u_h, u_h \rangle_{\mathcal{E}^\partial} - \langle [\![u_h]\!], \{a\nabla u_h\} \rangle_{\mathcal{E}} \\
& - \langle \frac{\tau^-}{\tau^+ + \tau^-} a^+ \nabla u_h^+ + \frac{\tau^+}{\tau^+ + \tau^-} a^- \nabla u_h^- - \frac{\tau^+\tau^-}{\tau^+ + \tau^-} [\![u_h]\!], [\![u_h]\!] \rangle_{\mathcal{E}^0} \\
& + \langle \frac{\tau^+ - \tau^-}{2(\tau^+ + \tau^-)} \mathbf{n}^+ [\![u_h]\!] - \frac{1}{\tau^+ + \tau^-} [\![a\nabla u_h]\!], [\![a\nabla u_h]\!] \rangle_{\mathcal{E}^0} \\
\geq\ & (a\nabla u_h, \nabla u_h)_{0,\mathcal{T}} + (du_h, u_h)_{0,\mathcal{T}} + \langle \tau_m [\![u_h]\!], [\![u_h]\!] \rangle_{\mathcal{E}} \\
& - \sum_{E \in \mathcal{E}} |\langle 3\{a\nabla u_h\}, [\![u_h]\!] \rangle_E| - \frac{1}{2} \sum_{E \in \mathcal{E}^0} |\langle \mathbf{n}^+ [\![u_h]\!], [\![a\nabla u_h]\!] \rangle_E| \\
& - \langle \frac{1}{\tau^+ + \tau^-} [\![a\nabla u_h]\!], [\![a\nabla u_h]\!] \rangle_{\mathcal{E}^0}
\end{aligned}
$$

Using (3.1) for the term on the last line and (3.4) and (3.3), we can further estimate

$$
\begin{aligned}
\mathcal{A}(u_h, u_h) \geq\ & (1 - (C_{3.4} + 6C_{3.3}) \frac{k+1}{2\tilde{\kappa}} - C_{3.1} \frac{k+1}{\tau^+ + \tau^-})(a\nabla u_h, \nabla u_h)_{0,\mathcal{T}} \\
& + \langle (\tau_m - 2\tilde{\kappa} h_{\mathcal{T}}^{-1})[\![u_h]\!], [\![u_h]\!] \rangle_{\mathcal{E}} \\
\geq\ & c_A |||v|||_h \ ,
\end{aligned}
$$

if we choose $\tilde{\kappa} \gtrsim (k+1)$ as well as γ large enough in $\tau = \gamma(k+1)^2 h_{\mathcal{T}}^{-1}$.　　　$\square$

Theorem 4.2.2

Let $(\mathbf{q}_h, u_h, \lambda_h) \in \mathbf{V}_h \times W_h \times M_h$ be the solution of the hybridized IPDG method with penalty parameter $\tau := \gamma(k+1)^j h_{\mathcal{T}}^{-1}$. Then, the consistency error estimators $\eta_{\mathcal{E},2}$ and $\eta_{\mathcal{E},3}$ of (4.1) can be bounded by

$$
\eta_{\mathcal{E},2} + \eta_{\mathcal{E},3} \leq C(\eta_{\mathcal{T},1} + \eta_{\mathcal{T},2} + \eta_{\mathcal{E},1}) \ ,
$$

where C depends on the local geometry of the underlying triangulation $\mathcal{T}$, if the constant γ in the penalty parameter $\tau = \gamma(k+1)^j h_{\mathcal{T}}^{-1}$, $j \geq 5/2$ is chosen big enough. For $1 \leq j < 5/2$, the constant $C = C(k)$ also depends on the polynomial degree k of W_h.

Proof. Let $v \in V^0 := W_k^c \cap C^0(\Omega)$ be continuous. We insert $v = (v - u_h) + v$ into (4.28) to get

$$
\mathcal{A}(u_h - v, u_h - v) = F(u_h - v) - \mathcal{A}(v, u_h - v). \tag{4.32}
$$

Let us focus on the second term on the right-hand side first. Using the fact that $v =$

$(v - u_h) + v$, we see that

$$
\begin{aligned}
\mathcal{A}(v, u_h - v) = {} & (-\mathbf{q}_h, \nabla(u_h - v))_{0,\mathcal{T}} + (du_h, u_h - v)_{0,\mathcal{T}} \\
& - \|a^{1/2}\nabla(u_h - v)\|_{0,\mathcal{T}}^2 - \|d^{1/2}(u_h - v)\|_{0,\mathcal{T}}^2 + (a\nabla u_h + \mathbf{q}_h, \nabla(u_h - v))_{0,\mathcal{T}} \\
& - \langle a\nabla v \cdot \mathbf{n} - \tau v, u_h - v \rangle_{\mathcal{E}^\partial} \\
& - \langle \frac{\tau^-}{\tau^+ + \tau^-}a^+\nabla v^+ + \frac{\tau^+}{\tau^+ + \tau^-}a^-\nabla v^-, [\![u_h]\!] \rangle_{\mathcal{E}^0} \\
& - \langle \frac{1}{\tau^+ + \tau^-}[\![a\nabla v]\!], [\![a\nabla(u_h - v)]\!] \rangle_{\mathcal{E}^0} \\
= {} & (\nabla \cdot \mathbf{q}_h + du_h, u_h - v)_{0,\mathcal{T}} + (a\nabla u_h + \mathbf{q}_h, \nabla(u_h - v))_{0,\mathcal{T}} \\
& - \|a^{1/2}\nabla(u_h - v)\|_{0,\mathcal{T}}^2 - \|d^{1/2}(u_h - v)\|_{0,\mathcal{T}}^2 + \langle \tau v - (\mathbf{q}_h + a\nabla v) \cdot \mathbf{n}, u_h - v \rangle_{\mathcal{E}^\partial} \\
& - \langle [\![\mathbf{q}_h]\!], \{u_h - v\} \rangle_{\mathcal{E}^0} - \langle \{\mathbf{q}_h\}, [\![u_h]\!] \rangle_{\mathcal{E}^0} \\
& - \langle \frac{\tau^- - \tau^+}{2(\tau^+ + \tau^-)}[\![a\nabla v]\!] \cdot \mathbf{n} - \{a\nabla v\}, [\![u_h]\!] \rangle_{\mathcal{E}^0} \\
& - \langle \frac{1}{\tau^+ + \tau^-}[\![a\nabla v]\!], [\![a\nabla(u_h - v)]\!] \rangle_{\mathcal{E}^0} \, .
\end{aligned}
$$

Inserting this result into equation (4.32), it follows that

$$
\begin{aligned}
\mathcal{A}(u_h - v, u_h - v) = {} & (f - \nabla \cdot \mathbf{q}_h - du_h, u_h - v)_{0,\mathcal{T}} + (a\nabla u_h + \mathbf{q}_h, \nabla(u_h - v))_{0,\mathcal{T}} \\
& + \|a^{1/2}\nabla(u_h - v)\|_{0,\mathcal{T}}^2 + \|d^{1/2}(u_h - v)\|_{0,\mathcal{T}}^2 \\
& + \langle [\![\mathbf{q}_h]\!], \{u_h - v\} \rangle_{\mathcal{E}^0} + \langle \{\mathbf{q}_h + a\nabla v\}, [\![u_h]\!] \rangle_{\mathcal{E}} \\
& + \langle \frac{\tau^- - \tau^+}{2(\tau^+ + \tau^-)}[\![a\nabla v]\!] \cdot \mathbf{n}, [\![u_h]\!] \rangle_{\mathcal{E}^0} \\
& + \langle \frac{1}{\tau^+ + \tau^-}[\![a\nabla v]\!], [\![a\nabla(u_h - v)]\!] \rangle_{\mathcal{E}^0} + \langle \tau(g_h - v), u_h - v \rangle_{\mathcal{E}^\partial} \\
& + \langle u_h - g_h, a\nabla(u_h - v) \cdot \mathbf{n} \rangle_{\mathcal{E}^\partial} \, .
\end{aligned}
\tag{4.33}
$$

Now, we choose the continuous interpolation $v := \pi_{\mathcal{T}} u_h \in W_k^c(\mathcal{T})$ of Remark 4.1.5 satisfying the Dirichlet boundary condtions. Using the Cauchy-Schwarz inequality and (4.23), we can bound the first two lines of the right hand side of (4.33) up to a constant, which only depends on the shape regularity of the triangulation, by

$$
\begin{aligned}
(\eta_{\mathcal{T},1} + \eta_{\mathcal{T},2}) &\|(k+1)h^{-1/2}[\![u_h]\!]\|_{\mathcal{E}_h} + \|(k+1)h^{-1/2}[\![u_h]\!]\|_{\mathcal{E}_h}^2 \\
& \lesssim (\eta_{\mathcal{T},1} + \eta_{\mathcal{T},2})\eta_{E,2} + (\eta_{E,2} + \eta_{E,3})^2 \, .
\end{aligned}
\tag{4.34}
$$

We use (3.2) and Remark (4.1.5) to estimate

$$\langle [\![\mathbf{q}_h]\!], \{u_h - v\}\rangle_{\mathcal{E}^0} \le \eta_{\mathcal{E},1}\frac{(k+1)^{1/2}}{h_{\mathcal{E}}^{1/2}}\{u_h - v\}\|_{0,\mathcal{E}^0} \lesssim \eta_{\mathcal{E},1}\frac{(k+1)^{3/2}}{h_T}(u_h - v)\|_{0,T}$$

$$\lesssim \eta_{\mathcal{E},1}(k+1)^{1/2}(\eta_{\mathcal{E},2} + \eta_{\mathcal{E},3}) \ . \tag{4.35}$$

On the other hand, we have

$$\langle \{\mathbf{q}_h + a\nabla v\}, [\![u_h]\!]\rangle_{\mathcal{E}} = \langle \{\mathbf{q}_h + a\nabla u_h\} + \{a\nabla(v - u_h)\}, [\![u_h]\!]\rangle_{\mathcal{E}}$$

$$\le \|\frac{h_{\mathcal{E}}^{1/2}}{(k+1)}(\{\mathbf{q}_h + a\nabla u_h\} + \{a\nabla(v - u_h)\})\|_{0,\mathcal{E}}(\eta_{\mathcal{E},2} + \eta_{\mathcal{E},3})$$

$$\lesssim (\|\mathbf{q}_h + a\nabla u_h\|_{0,T} + \|a\nabla(v - u_h)\|_{0,T})\,(\eta_{\mathcal{E},2} + \eta_{\mathcal{E},3})$$

$$\lesssim (\eta_{T,1} + \eta_{\mathcal{E},2} + \eta_{\mathcal{E},3})\,(\eta_{\mathcal{E},2} + \eta_{\mathcal{E},3}) \ . \tag{4.36}$$

Adding $0 = [\![a\nabla u_h + \mathbf{q}_h]\!] - [\![a\nabla u_h + \mathbf{q}_h]\!]$ to the following term, we get

$$\langle [\![a\nabla v]\!], \frac{[\![a\nabla(u_h - v)]\!]}{\tau^+ + \tau^-}\rangle_{\mathcal{E}^0} \le \langle [\![\mathbf{q}_h + a\nabla u_h]\!] - [\![\mathbf{q}_h]\!], \frac{[\![a\nabla(u_h - v)]\!]}{\tau^+ + \tau^-}\rangle_{\mathcal{E}} - \|\frac{[\![a\nabla(u_h - v)]\!]}{\sqrt{\tau^+ + \tau^-}}\|_{0,\mathcal{E}}^2$$

$$\lesssim ((k+1)\eta_{T,2} + (k+1)^{1/2}\eta_{E,1})\|h_E^{-1/2}\frac{[\![a\nabla(u_h - v)]\!]}{\tau^+ + \tau^-}\|_{0,E} - 0$$

$$\lesssim (\eta_{T,2} + \eta_{E,1})\eta_{E,2} \ , \tag{4.37}$$

where we have used $\tau = \gamma(k+1)^2 h_T^{-1}$. Similarly, we estimate

$$\langle \frac{\tau^- - \tau^+}{2(\tau^+ + \tau^-)}[\![a\nabla v]\!] \cdot \mathbf{n}, [\![u_h]\!]\rangle_{\mathcal{E}^0} \le \langle \frac{\tau^- - \tau^+}{2(\tau^+ + \tau^-)}([\![a\nabla(v - u_h)]\!] + [\![a\nabla u_h + \mathbf{q}_h]\!] - [\![\mathbf{q}_h]\!]), [\![u_h]\!] \cdot \mathbf{n}\rangle_{\mathcal{E}^0}$$

$$\le \|a\nabla(v - u_h)\|_{0,T} + \|a\nabla u_h + \mathbf{q}_h\|_{0,T} + (k+1)^{-1/2}\eta_{E,1})\eta_{E,2}$$

$$\lesssim ((\eta_{\mathcal{E},2} + \eta_{\mathcal{E},3}) + \eta_{T,1} + \eta_{E,1})\eta_{E,2} \ . \tag{4.38}$$

By the choice of v, we have $v = \pi_T u_h = g_h$ on $\partial\Omega$. Hence, the boundary terms can be estimated by

$$\langle g_h - v, u_h - v\rangle_{\mathcal{E}^\partial} + \langle u_h - g_h, a\nabla(u_h - v)\cdot\mathbf{n}\rangle_{\mathcal{E}^\partial} \le \eta_{\mathcal{E},3}\|\frac{h_{\mathcal{E}}^{1/2}}{k+1}a\nabla(u_h - v)\cdot\mathbf{n}\|_{0,\mathcal{E}^\partial}$$

$$\lesssim \eta_{\mathcal{E},3}\|a\nabla(u_h - v)\|_{0,T}$$

$$\lesssim \eta_{\mathcal{E},3}(\eta_{\mathcal{E},2} + \eta_{\mathcal{E},3}) \ . \tag{4.39}$$

Since $[\![v]\!] = v$ on boundary faces, we have $\|u_h - g_D\|_{\mathcal{E}^\partial} \le \|[\![u_h]\!]\|_{\mathcal{E}^\partial}$. Thus we use estimates

(4.34), (4.35), (4.36), (4.37), (4.38), (4.39) in (4.33) to arrive at

$$
\begin{aligned}
\gamma(\eta_{\mathcal{E},2} + \eta_{\mathcal{E},3})^2 &\leq \gamma \| (k+1) h_{\mathcal{E}}^{-1/2} [\![u_h]\!] \|_{\mathcal{E}}^2 \\
&\leq \| u_h - v \|^2 \leq c_A^{-1} \mathcal{A}(u_h - v, u_h - v) \\
&\leq C\big((\eta_{\mathcal{T},1} + \eta_{\mathcal{T},2} + \eta_{\mathcal{E},1}(k+1)^{1/2})(\eta_{\mathcal{E},2} + \eta_{\mathcal{E},3}) + (\eta_{\mathcal{E},2} + \eta_{\mathcal{E},3})^2 \big) \, ,
\end{aligned}
$$

where the constant C depends only on the shape regularity of the triangulation.

After rearranging this inequality, we see that we obtain the desired estimate, if $\gamma \geq \sqrt{k+1}$ is chosen large enough. $\square$

For the hybridized LDG methods the jumps in the primal variable admit the following property:

Theorem 4.2.3

Let $(\mathbf{q}_h, u_h, \lambda_h) \in \mathbf{V}_h \times W_h \times M_h$ be the solution of any of the hybridized LDG methods of Table 3.2. Then, the consistency error estimators $\eta_{\mathcal{E},2}$ and $\eta_{\mathcal{E},3}$ of (4.1) can be bounded by

$$
\eta_{\mathcal{E},2} + \eta_{\mathcal{E},3} \leq C(\eta_{\mathcal{T},1} + \eta_{\mathcal{T},2} + \eta_{\mathcal{E},1}) \, ,
$$

where C depends on the local geometry of the underlying triangulation $\mathcal{T}$, if the constant γ in the penalty parameter $\tau = \gamma(k+1)^j h_{\mathcal{T}}^{-1}$, $j \geq 5/2$ is chosen big enough. For $0 \leq j < 5/2$, the constant $C = C(k)$ also depends on the polynomial degree k of W_h.

Proof. We start with equation (3.8b), the second characterizing equation for the H-DG methods. From equation (3.27) we know that for the LDG method on interior faces, we have

$$
\hat{\mathbf{q}}_h = \frac{\tau^-}{\tau^- + \tau^+} \mathbf{q}_h^+ + \frac{\tau^+}{\tau^- + \tau^+} \mathbf{q}_h^- + \frac{\tau^+ \tau^-}{\tau^- + \tau^+} [\![u_h]\!] \, . \tag{4.40}
$$

On exterior faces, we see directly from the definitions that

$$
\begin{aligned}
\hat{\mathbf{q}}_h &= \hat{\mathbf{Q}} \lambda_h + \hat{\mathbf{Q}} g_h + \hat{\mathbf{Q}} f \\
&= (\mathbf{Q} \lambda_h + \mathbf{Q} g_h + \mathbf{Q} f) + \tau(U \lambda_h + U g_h + U f - \lambda_h - g_h)\mathbf{n} \\
&= \mathbf{q}_h + \tau(u_h - g_h)\mathbf{n} \, , \tag{4.41}
\end{aligned}
$$

since $\lambda_h = 0$ on boundary faces.

From (4.16) we know that (3.8b) can be written as

$$\langle \frac{\tau^+ \tau^-}{\tau^+ + \tau^-}[\![u_h]\!], [\![w_h]\!]\rangle_{\mathcal{E}^0} + \langle \tau(u_h - g_h), w_h\rangle_{\mathcal{E}^\partial} =$$

$$= (f - du_h - \nabla\!\cdot\!\mathbf{q}_h, w_h)_{0,\mathcal{T}} + \langle [\![\mathbf{q}_h]\!], \{w_h\} + \frac{\tau^+ - \tau^-}{\tau^+ + \tau^-}\mathbf{n}^+[\![w_h]\!]\rangle_{\mathcal{E}^0} \quad (4.42)$$

after realizing that $\hat{\mathbf{q}}_h$ is single valued on interior edges and using integration by parts.

We use the continuous interpolation $\pi_{\mathcal{T}} u_h \in W_h(\mathcal{T}) \cap H_0^1(\Omega)$ of Lemma 4.1.4 to choose a suitable test function $w := u_h - \pi_{\mathcal{T}} u_h$. Since $\pi_{\mathcal{T}} u_h$ is continuous, we know that $[\![u_h - \pi_{\mathcal{T}} u_h]\!] = [\![u_h]\!]$ on interior faces. Using $\tau_m = \min\{\tau^+, \tau^-\} \leq \frac{\tau^+ \tau^-}{\tau^+ + \tau^-}$ we deduce the estimate

$$\|\tau_m^{1/2}[\![u_h]\!]\|_{0,\mathcal{E}^0}^2 + \|\tau^{1/2}(u_h - g_h)\|_{0,E}^2$$

$$\leq |\frac{h_{\mathcal{T}}}{k+1}(f - \nabla\!\cdot\!\mathbf{q}_h - du_h)\|_{0,\mathcal{T}}\|(k+1)h_{\mathcal{T}}^{-1}w\|_{\mathcal{T}}$$

$$+ |\frac{h_{\mathcal{E}}^{1/2}}{(k+1)^{1/2}}[\![\mathbf{q}_h]\!]\|_{0,\mathcal{E}}(k+1)^{1/2}\left(\|h_{\mathcal{E}}^{-1/2}\{w\}\|_{0,\mathcal{E}_h^0} + \|h_{\mathcal{E}}^{-1/2}\frac{\tau^+ - \tau^-}{\tau^+ + \tau^-}[\![w]\!]\|_{0,\mathcal{E}_h^0}\right)$$

$$\lesssim \left(\eta_{\mathcal{T},2} + \eta_{\mathcal{E},1}(k+1)^{1/2}\right)\|(k+1)h_{\mathcal{T}}^{-1}w\|_{0,\mathcal{T}} . \quad (4.43)$$

Using Lemma 4.1.4, for $\tau = \gamma h_{\mathcal{E}}^{-1}(k+1)^j$, where $0 < \gamma$ is chosen sufficiently large but independent of k and h_E, this implies:

$$(\|(k+1)h_{\mathcal{E}}^{-1/2}[\![u_h]\!]\|_{0,\mathcal{E}^0} + \|(k+1)h_{\mathcal{E}}^{-1/2}(u_h - g_h)\|_{0,\mathcal{E}^\partial})^2$$

$$\lesssim (k+1)^{5/2-j}\left(\eta_{\mathcal{T},2} + \eta_{\mathcal{E},1}\right)\left(\|(k+1)h_{\mathcal{E}}^{-1/2}[\![u_h]\!]\|_{0,\mathcal{E}_h^0} + \|(k+1)h_{\mathcal{E}}^{-1/2}(u_h - g_h)\|_{0,\mathcal{E}_h^\partial}\right)$$

Dividing both sides of the inequality by $\|(k+1)h_{\mathcal{E}}^{-1/2}[\![u_h]\!]\|_{0,\mathcal{E}_h^0} + \|(k+1)h_{\mathcal{E}}^{-1/2}(u_h - g_h)\|_{0,\mathcal{E}_h^\partial}$, we get

$$(\sum_{E\in\mathcal{E}_h^0}\eta_{E,2}^2)^{1/2} + (\sum_{E\in\mathcal{E}_h^\partial}\eta_{E,3}^2)^{1/2} = \|(k+1)h_{\mathcal{E}}^{-1/2}[\![u_h]\!]\|_{0,\mathcal{E}_h^0} + \|(k+1)h_{\mathcal{E}}^{-1/2}(u_h - g_h)\|_{0,\mathcal{E}_h^\partial} \leq$$

$$\leq (k+1)^{5/2-j}(\eta_{\mathcal{T},2} + \eta_{\mathcal{E},1}) . \quad (4.44)$$

Consequently, for $j = 5/2$ we obtain the strict upper bound

$$(\sum_{E\in\mathcal{E}_h^0}\eta_{E,2}^2)^{1/2} + (\sum_{E\in\mathcal{E}_h^\partial}\eta_{E,3}^2)^{1/2} \leq \eta_{\mathcal{T},2} + \eta_{\mathcal{E},1} , \quad (4.45)$$

while for other values of $0 < j \leq 5/2$ the consistency error $\eta_{\mathcal{E},2} + \eta_{\mathcal{E},3}$ is only bounded up to a constant depending on k. □

4.3 Estimation of the jump term for the H-NC methods

For the hybridized discontinuous Galerkin methods, $\eta_{E,2}$ and $\eta_{E,3}$ measuring the error induced by the jumps in the primal variable u_h can be bounded by the remaining estimator terms. For the nonconforming methods we can not hope to get an equivalent result. Since $-a\nabla u_h^{NC} = \mathbf{q}_h^{NC}$, for element-wise constant data a and f we know a priori that $\eta_{T,1} = \eta_{E,\nabla} = 0$. Furthermore, it has been shown in [30] that for piecewise constant f and $d = 0$ there holds $\mathbf{q}_h^{RT}(x) = -a\nabla u_h^{NC}(x) + \frac{f}{2}(x - x_T)$ for all $x \in T \in \mathcal{T}$ with barycenter x_T, if we use the lowest order approximations. The main idea of the connection between the nonconforming approximation u_h^{NC} and $\mathbf{q}_h^{RT}$ dates back to [4]. This implies that for the nonconforming method $\eta_{E,1}$ vanishes for $f = 0$ as well, since $\mathbf{q}_h^{RT}$ is $H(div)$-conforming.

Now, we discretize the stationary diffusion equation $-\nabla \cdot a\nabla u = 0$ in Ω with vanishing right-hand side and inhomogeneous Dirichlet boundary data using the lowest order (hybridized) nonconforming method. Since $\nabla \cdot \mathbf{q}_h = 0$, we see that $\eta_{T,2}$ vanishes as well. So in the end there is no estimator term left, that could give an upper bound for the jumps present in the primal variable.

Still we can recover the residual type estimator of [19] which has been proven to be efficient. We are going to show this in the sequel.

Lemma 4.3.1

The primal variable of the H-NC method of odd order k is continuous across the faces of $\mathcal{T}$ at the $k - 1$ roots of the Legendre polynomial of degree k, in particular at the midpoints of the faces.

Proof. From equation (3.56) we know that on any interior face $E \in \mathcal{E}^0$ the jump of the primal variable u, which we denote by $p_J := u^+ - u^-$, satisfies

$$\langle p_J, p\rangle_E = 0 \quad \forall p \in \mathbb{P}_k(E).$$

Expressing p_J in the $L_2(E)$-orthogonal basis of the Legendre polynomials $p_i \in \mathbb{P}_k(E)$, $0 \leq i \leq k$ we get

$$0 = \langle p_J, p\rangle_E = \sum_{i=1}^{k} \alpha_i \langle p_i, p\rangle_E = \alpha_k \langle p_k, p\rangle_E.$$

Hence, we see that p_J is a multiple of the k-th Legendre polynomial. Therefore, $p_J(x) = 0$ for x being a root of the k-th Legendre polynomial which means that u_h is continuous at these points. $\square$

The continuity of u_h at k points on each face can be exploited to prove the following lemma.

Lemma 4.3.2
For the H-NC methods, $\eta_{E,2}$ can be bounded by

$$\eta_{E,2} \leq \eta_{E,t} := \frac{1}{\sqrt{2}} h_E^{1/2} \| \frac{\partial}{\partial t}(u_h^+ - u_h^-)\|_{0,E} \ ,$$

and equivalently, $\eta_{E,3}$ can be bounded by

$$\eta_{E,t\partial} := h_E^{1/2} \| \frac{\partial}{\partial t}(u_h^- - g_D)\|_{0,E} \ ,$$

where $\frac{\partial}{\partial t} u$ denotes the directional derivative in the tangential direction t along the edge E.

Proof. The proof is easily established by bounding $u_h^+ - u_h^-$ which we abbreviate with p_J again. Since at the midpoint of the face E we have $p_J(x_M) = 0$, with $E' := [x_M, x]$ we may estimate

$$p_J(x) = (p_J(x) - p_J(x_M)) = \int_{x_M}^x \frac{\partial}{\partial t} p_J(x(t)) dt = \langle 1, \frac{\partial}{\partial t} p_J \rangle_{E'} \leq |E'|^{1/2} \| \frac{\partial}{\partial t} p_J\|_{E'}$$

using the fundamental theorem of calculus and the Cauchy-Schwarz inequality. Since $|E'| \leq \frac{1}{2}|E| = \frac{1}{2}h_E$, this can be used to obtain

$$\|p_J\|_E^2 = \int_E p_J(x(t))^2 dt \leq \int_E |E'| \|\frac{\partial}{\partial t} p_J\|_{E'}^2 dt \leq \frac{1}{2} h_E^2 \|\frac{\partial}{\partial t} p_J\|_E^2 \ .$$

This gives the desired result. For the boundary term one proceeds exactly the same way. $\square$

Remark 4.3.3
One may hope to get an additional factor $\sqrt{1/(k+1)}$ instead of $1/2$ if choosing a suitable root of p_J for higher order approximations. Unfortunately, the roots of the Legendre polynomials are not distributed uniformly on the interval. So we omit the technical and marginal modification that involves the search for the maximal distance of two neighboring roots of the Legendre polynomial of degree k.

4.4 Efficiency of the error estimator

In this section we show that the error estimator η gives not only an upper bound to the actual error

$$e_h := (\sum_{T \in \mathcal{T}} e_h(T))^{1/2}, \text{ where} \tag{4.46}$$
$$e_h(T) := (\|u - u_h\|_{1,T}^2 + \|\mathbf{q} - \mathbf{q}_h\|_{0,T}^2)^{1/2}$$

but also a local lower bound up to constants and data oscillation.

Theorem 4.4.1
Let $(\mathbf{q}_h, u_h, \lambda_h) \in \mathbf{V}_h \times W_h \times M_h$ be the solution of the hybridized Discontinuous Galerkin methods of table 3.2. Then the a posteriori error estimator η defined in (4.1) can be bounded by the error with respect to the solution of the primal mixed formulation (2.4) $(\mathbf{q}, u) \in \mathbf{V} \times W$, where $\mathbf{V} := (L^2(\Omega))^2$ and $W := H^1(\Omega)$ up to a constant $C(k)$ that may depend on the polynomial degree k of the discrete space W_h and data oscillation.

$$\eta(T)^2 \leq C(k)(e_h^2 + osc_f(T)) , \tag{4.47}$$

where $osc_f(D) := \|f - f_h\|_{0,D}, \ D \subset \Omega$.

An estimate of this kind guarantees, that the quantity on the right hand side – the true error of our computed solution – can be bounded from below by the error estimator η up to constants that depend only on the local geometry of the triangulation $\mathcal{T}$ and the polynomial degree k of the approximation space.

If Estimate (4.47) did not hold true the error could decrease faster than the estimator during the adaptive cycle. So the error might reach a prescribed tolerance significantly earlier than the estimator. So we would spend unnecessary work to achieve some prescribed error tolerance since we can only control the estimated error directly. For this reason we call an error estimator efficient if it satisfies an estimate like (4.47).

In order to prove estimate (4.47) we make use of so called bubble functions b_T, b_E, sufficiently regular functions with local support.

Definition 4.4.2 (bubble functions)
For an element $T \in \mathcal{T}$ and a patch ω_E consisting of the two elements adjacent to edge $E \in \mathcal{E}$ we define the element bubble function $b_T \in C^1(T)$ and edge bubble function $b_E \in C^1(w_E)$

as

$$b_T(x) \quad := 27\lambda_1(x)\lambda_2(x)\lambda_3(x)$$
$$b_E(x)|_T := 4\lambda_1(x)\lambda_2(x), \ T \in \{T^+, T^-\}$$

where $\lambda_1(x)$, $\lambda_2(x)$ are the barycentric coordinates associated with the endpoints of edge E in element T and $\lambda_3(x)$ is the remaining barycentric coordinate of T.

In the following me use the following properties of b_T and b_E:

Proposition 4.4.3 (properties of bubble functions)
Let $p \in \mathbb{P}_k(T)$ be a polynomial of degree k on T, $p \in \mathbb{P}_k(E)$ respectively. Then we have the following estimates in connection with the bubbles b_T and b_E:

$$(p, pb_T)_T \leq (p, p)_T \leq C_T(k)(p, pb_T)_T \ , \tag{4.48}$$

$$\langle p, pb_E \rangle_E \leq \langle p, p \rangle_E \leq C_E(k)\langle p, pb_E \rangle_E \ . \tag{4.49}$$

Proof. The left inequality is a direct consequence of $\max_{x \in \omega_E}(b_E(x)) = 1$ and $\max_{x \in T}(b_T(x)) = 1$. The right hand side inequality can be easily seen after transforming the integrals to the reference element using the positivity of the bubbles to define a weighted L_2-norm and using the norm equivalence on finite dimensional vector spaces.

To get explicit bounds for the constants $C_T(k)$ and $C_E(k)$ we have to invest a bit more work. We use a quadrature formula based on Gaussian quadrature to calculate the integral on the right hand side of Equation (4.48). Since we only have interior quadrature points, we can estimate this integral from below by the minimal value of the bubble among all quadrature points and the integral over p^2. This leads to explicit expressions of the desired constants $C_E(k) = \frac{1}{4}(k+2)^2$ for $k \leq 100$ and $C_T(k) = 35(k+3)^3$ for $k \leq 10$. □

These estimates are competitive with the general estimates $C_T(k) := (2k+2)^{2d}(\frac{d}{d+1})^{d+1}d!$ derived in [56], where d is the dimension of the simplex $T \subset \Omega \in \mathbb{R}^d$.

Furthermore we use the following estimates:

Proposition 4.4.4
Let $p \in \mathbb{P}_k(T)$ for some $T \in \mathcal{T}$, $k \in \mathbb{N}$. Then we have

$$\|\nabla(pb_T)\|_{0,T} \lesssim \gamma_1(k)h_T^{-1}\|p\|_{0,T} \ , \tag{4.50}$$

$$\|\sigma b_E\|_{0,T} \lesssim \gamma_2 h_E^{1/2}\|\sigma\|_{0,E} \ , \tag{4.51}$$

$$\|\nabla(\sigma b_E)\|_{0,T} \lesssim \gamma_3(k)h_E^{-1/2}\|\sigma\|_{0,E} \ , \tag{4.52}$$

where $\gamma_1(k) \leq \frac{27}{4}(k+3)$, $\gamma_2 \leq \frac{1}{2}$ and $\gamma_3(k) \leq 6(k+2)$.

The proof of these estimates in [56] makes use of orthogonal polynomials on the reference elements.

Proof of Theorem 4.4.1. For the hybridized DG methods of table 3.2 we know that $\eta_{E,2}$ and $\eta_{E,3}$ can be bounded by the other estimator terms. Thus it suffices to estimate (a) $\eta_{T,1}$, (b) $\eta_{T,2}$ and (c) $\eta_{E,1}$.

(a) For this estimate we use the fact that $\mathbf{v}_1 := (a^{-1}\mathbf{q}_h + \nabla u_h)|_T \in L^2(\Omega)$ and $\mathbf{v}_2 := (a^{-1}(\mathbf{q}_h - \mathbf{q}) + \nabla(u_h - u))|_T \in L^2(\Omega)$ are admissible test-functions in (2.4a) to get

$$
\begin{aligned}
\eta_{T,1}^2 &= (a^{-1}\mathbf{q}_h + \nabla u_h, \mathbf{v}_1)_{0,\mathcal{T}} = (a^{-1}(\mathbf{q}_h - q) + \nabla(u_h - u), \mathbf{v}_1)_{0,T} \\
&= (\mathbf{v}_2, a^{-1}(\mathbf{q}_h - \mathbf{q}) + \nabla(u_h - u))_{0,T} = \|a^{-1}(\mathbf{q}_h - \mathbf{q}) + \nabla(u_h - u)\|_{0,T}^2 \\
&\leq 2(\|a^{-1}(\mathbf{q} - \mathbf{q}_h)\|_{0,T}^2 + \|\nabla(u - u_h)\|_{0,T}^2)
\end{aligned}
\tag{4.53}
$$

(b) We start to estimate the L^2-norm in $\eta_{T,2} := \frac{h_T}{k+1}\|f_h - \nabla\cdot\mathbf{q}_h - du_h\|_{0,T}$, where we make use of the properties of the element bubble in the first estimate.

$$
\Upsilon := \|f_h - \nabla\cdot\mathbf{q}_h - du_h\|_{0,T}^2 \lesssim C_T(k)(f_h - \nabla\cdot\mathbf{q}_h - du_h, (f_h - \nabla\cdot\mathbf{q}_h - du_h)b_T)_T \ .
\tag{4.54}
$$

Now $v_T := (f_h - \nabla\cdot\mathbf{q}_h - du_h)b_T \in C^0(\Omega)$ vanishing on $\Omega \setminus \overset{o}{T}$ is an admissible test-functions in the variational formulation of the pde. So we add $0 = f - f$ inside the inner product use integration by parts and equation (2.4b) to rearrange the terms

$$
\begin{aligned}
\Upsilon &\lesssim C_T(k)((f - \nabla\cdot\mathbf{q}_h - du_h, v_T)_\mathcal{T} + (f_h - f, v_T)_T) \\
&= C_T(k)((f, v_T)_T - (du_h, v_T)_T + (\mathbf{q}_h, \nabla v_T)_T + (f_h - f, v_T)_T) \\
&= C_T(k)(d(u - u_h), v_T)_T + (\mathbf{q}_h - \mathbf{q}, \nabla v_T)_T + (f_h - f, v_T)_T) \ .
\end{aligned}
$$

Using the Cauchy-Schwarz inequality and estimates (4.48) and (4.50) for the bubble functions we further estimate

$$
\begin{aligned}
\Upsilon &\leq C_T(k)(\|d(u - u_h)\|_{0,T}\|v_T\|_{0,T} + \|\mathbf{q}_h - \mathbf{q}\|_{0,T}\|\nabla v_T\|_{0,T} + \|f_h - f\|_{0,T}\|v_T\|_{0,T}) \\
&\lesssim C_T(k)(\|d(u - u_h)\|_{0,T} + \gamma_1(k)h_T^{-1}\|\mathbf{q}_h - \mathbf{q}\|_{0,T} + \|f_h - f\|_{0,T})\Upsilon^{1/2}
\end{aligned}
$$

Rearranging the overall inequality we conclude

$$\eta_{T,2} \lesssim \frac{h_T C_T(k) \max\{\gamma_1(k)h_T^{-1}, 1\}}{k+1}(e_h + osc_f(\mathcal{T})) \ ,$$

which – using the explicit bounds for the constants $\gamma_1(k), C_T(k)$ with h_T small enough – can be simplified to

$$\eta_{T,2} \lesssim 709(k+3)^3(e_h + osc_f(\mathcal{T})) \ . \tag{4.55}$$

(c) For the estimation of $\eta_{E,1} := \sqrt{\frac{h_E}{k+1}}\|[\![\mathbf{q}_h]\!]\|_{0,E}$ we add the discrete solutions $-u_h$ and $\mathbf{q}_h$ to both sides of (2.4b) to get

$$(\mathbf{q}_h - \mathbf{q}, \nabla w) + (d(u - u_h), w) = (f, w) + (\mathbf{q}_h, \nabla w) - (du_h, w) \ \forall w \in H_0^1(\Omega) \ . \tag{4.56}$$

Since $[\![\mathbf{q}_h]\!]$ is only defined on $\mathcal{E}_h$ we define some suitable polynomial extension

$$p_j \in W_h : p_j|_E = [\![\mathbf{q}]\!] \ .$$

It is easy to see that $v_E := p_j b_E \in H_0^1(\omega_E)$ is an admissible test-functions for equation (4.56) for any interior edge $E \in \mathcal{E}^o$. So we test (4.56) with v_E and use integration by parts on T^+ and T^- to get

$$(\mathbf{q}_h - \mathbf{q}, \nabla v_E) + (d(u - u_h), v_E)$$
$$= (f, v_E) - (\nabla\cdot\mathbf{q}_h, v_E) + \langle[\![\mathbf{q}_h]\!], v_E\rangle_E - (du_h, v_E) \ . \tag{4.57}$$

So we estimate

$$\|[\![\mathbf{q}_h]\!]\|_{0,E}^2 = \langle[\![\mathbf{q}_h]\!], [\![\mathbf{q}_h]\!]\rangle_E \lesssim C_E(k)\langle[\![\mathbf{q}_h]\!], \overbrace{[\![\mathbf{q}_h]\!]b_E}^{=:v_E}\rangle_E$$
$$\lesssim C_E(k)\left((\mathbf{q}_h - \mathbf{q}, \nabla v_E)_{\omega_E} + (d(u - u_h), v_E)_{\omega_E} - (f - \nabla\cdot\mathbf{q}_h - du_h, v_E)_{\omega_E}\right)$$
$$= C_E(k)\left(\gamma_3(k)h_E^{-1/2}\|\mathbf{q}_h - \mathbf{q}\|_{0,\omega_E} + \gamma_2 h_E^{1/2}\|d(u - u_h)\|_{0,\omega_E} + \right.$$
$$\left. + \gamma_2 h_E^{1/2}\|f - \nabla\cdot\mathbf{q}_h - du_h\|_{0,\omega_E}\right)\|[\![\mathbf{q}_h]\!]\|_{0,\omega_E}$$
$$\lesssim C_E(k)(\frac{\gamma_3(k)}{\sqrt{h_E}}e_{h,\omega_E} + \gamma_2 h_E^{1/2}\|f - \nabla\cdot q_h - du_h\|_{0,\omega_E})\|[\![\mathbf{q}_h]\!]\|_{0,\omega_E} \tag{4.58}$$

using the properties of the edge bubbles, equation (4.57), the Cauchy-Schwarz inequality, estimates (4.51) and (4.52) for the edge bubbles and the fact that $\gamma_2\sqrt{h_E} \leq \gamma_3(k)$ on a sufficiently fine mesh. Thus we get the estimate

$$
\begin{aligned}
\eta_{E,1} &\lesssim \frac{h_E^{1/2}}{(k+1)^{1/2}}C_E(k)(\frac{\gamma_3(k)}{\sqrt{h_E}}e_{h,\omega_E} + \gamma_2 h_E^{1/2}\|f - \nabla\cdot q_h - du_h\|_{0,\omega_E}) \\
&= \frac{C_E(k)\gamma_3(k)}{(k+1)^{1/2}}e_{h,\omega_E} + \frac{C_E(k)\gamma_2}{(k+1)^{1/2}}h_E\|f - \nabla\cdot q_h - du_h\|_{0,\omega_E}) \\
&\approx \frac{C_E(k)\gamma_3(k)}{(k+1)^{1/2}}e_{h,\omega_E} + C_E(k)\gamma_2(k+1)^{1/2}\sum_{T^+,T^-}(\eta_{T,2} + \frac{h_E}{k+1}osc_T(f)) \\
&\lesssim 3(k+2)^{3.5}e_{h,\omega_E} + 89\sum_{T^+,T^-}((k+3)^{5.5}e_{h,T} + h_E 3(k+3)^{4.5}osc_T(f)) \\
&\lesssim 90(k+3)^{5.5}e_{h,\omega_E} + h_E 270(k+3)^{4.5}osc_{\omega_E}(f) .
\end{aligned}
\tag{4.59}
$$

Due to the fact that the estimation of $\eta_{E,1}$ involves $\eta_{T,2}$ again, the constants depending on the polynomial degree k had to be multiplied with the constant of the estimation of $\eta_{T,2}$. □

Chapter 5

Numerical results

5.1 Numerical solution of hybrid methods

All numerical tests that are documented in the following chapter are performed in Matlab. For the hybrid methods, using polynomial spaces $\mathbf{V}_h$, W_h, which consist of full polynomials on each element $T \in \mathcal{T}$, the code is based on the Matlab library of nodal unstructured discontinuous Galerkin methods `nudg`, which is extensively documented in [33]. The hybridized Raviart-Thomas method has been implemented using nodal bases for W_h and M_h and a modal basis for $\mathbf{V}_h$, such that $l_j^{RT}(\phi_i) = \delta_{ij}$, $1 \leq i, j \leq dim(\mathbf{V}(T))$ for each basis function ϕ_i of $\mathbf{V}(T)$, where l_j^{RT} denotes the j-th degree of freedom of the Raviart-Thomas method on $T \in \mathcal{T}$, see Remark 3.18.

For the practical application of the hybrid methods discussed so far we implement the standard cycle of Adaptive Finite Element Methods consisting of the modules

$$\hookrightarrow \text{SOLVE} \rightarrow \text{ESTIMATE} \rightarrow \text{MARK} \rightarrow \text{REFINE} ,$$

which is repeated until the estimator η reaches some prescribed error tolerance `TOL`. Given a triangulation $\mathcal{T}$, the spaces $\mathbf{V}_h$, W_h, M_h, and the dependence of the numerical flux $\hat{q}_h$ on $\mathbf{q}_h, u_h, \hat{u}_h, \tau$, the module SOLVE consists of three steps. The first step involves the construction of discrete local mappings

$$(\mathbf{Q}(\cdot), \mathbf{U}(\cdot)) : \ M(\partial T) \rightarrow \mathbf{V}(T) \times W(T), \ m \mapsto (\mathbf{q}_m, u_m) ,$$

and

$$(\mathbf{Q}(\cdot), \mathbf{U}(\cdot)) : \ L^2(T) \rightarrow \mathbf{V}(T) \times W(T), \ f \mapsto (\mathbf{q}_f, u_f) ,$$

defined by the linear systems (3.5) and (3.6). After choosing a suitable basis, these local mappings can be represented by matrices Q_m^{qu} and Q_f^{qu}, respectively. The second part consists of the solution of the global linear system

$$a_h(\eta, \mu) = b(\mu) \qquad\qquad \forall \mu \in M_h ,$$

where the bilinear form $a_h : M_h \times M_h \to \mathbb{R}$ is chosen from Table 3.2, according to the hybrid method under consideration. To solve this equation, we determine the coefficients α_i, $1 \le dim(M_h)$ of $\eta_h = \sum_{1 \le i \le dim(M_h)} \alpha_i \breve{l}_i$, where $\breve{l}_i$ denotes the i-th basis functions of M_h. To get the matrix representation $\mathfrak{A}$ of the bilinear form $a_h(\cdot, \cdot)$, defined by $\mathfrak{A}_{i,j} = a_h(\breve{l}_j, \breve{l}_i)$ we use the matrices Q_m^{qu} and Q_f^{qu} of the local mappings. In the third step we recover the solution of the hybrid method using the local mappings

$$(\mathbf{q}_h, u_h, \lambda_h) := (\mathbf{Q}f + \mathbf{Q}\eta + \mathbf{Q}g_D, \mathbf{U}f + \mathbf{U}\eta + \mathbf{U}g_D, \eta + g_D) .$$

The module ESTIMATE of the adaptive algorithm consists of the computation of the a posteriori error estimator η of Chapter 4 and its local distribution on $\mathcal{T}$.

If the estimator η has not reached the prescribed error tolerance TOL, the module MARK is used to select elements for refinement based on the local distribution of the error estimator. As marking strategy we use a bulk criterion also known as Dörfler's marking introduced in [31]. Therefore, we define $\eta_T^2 := \eta_{T,1}^2 + \frac{1}{2} \sum_{E \subset \partial T} (\eta_{E,1}^2 + \eta_{E,2}^2) + \eta_{E,3}^2$ for all $T \in \mathcal{T}$ and sort the elements $T \in \mathcal{T}$ according to η_T in descending order $\mathcal{T}^s := \{T_1, \ldots, T_{|\mathcal{T}|}\}$ using a greedy algorithm. On the basis of this list, we determine the minimal index $m \in \mathbb{N}$, such that the contribution of the error estimator from the subset $\mathcal{T}^m := \{T_1, \ldots, T_m\} \subset \mathcal{T}^s$ exceeds a certain fractional part of the total error estimator η, i.e.

$$\sum_{T \in \mathcal{T}^m} \eta_T^2 > \theta \eta^2,$$

where $0 < \theta < 1$.

Finally, the module REFINE is used to refine the triangulation $\mathcal{T}$ in such a way that the local mesh size h_T of each marked element T is halved and shape-regularity of the conforming triangulation is preserved. Therefore, we use newest vertex bisection, where each marked element is bisected twice.

For newest vertex bisection, in each element one edge is declared as a refinement edge. On the initial triangulation, the refinement edges are declared a priori in a suitable way. For instance, the longest edge in each element may be chosen as the refinement edge. In an element which has been created by refinement of the triangulation, the edge opposite to the newest created vertex is declared as refinement edge. In order to ensure conformity of the triangulation, an element is refined only if the refinement edge of the element to be refined is also the refinement edge of the adjacent element or if it is a boundary edge. If the refinement edge of the adjacent element is not yet compatible, this neighboring element has to be refined first. An extensive documentation of this refinement strategy can be found in [40].

We test the hybrid methods with a variety of test problems. In order to confirm the well-known a priori estimates of Chapter 3 we introduce the experimental order of convergence (EOC) of the error $\|u - u_h\|_{1,\mathcal{T}}$ in terms of the globally coupled degrees of freedom (DOFS), defined by

$$EOC_{h,1} := \log\left(\frac{\|u - u_{2h}\|_{1,\mathcal{T}}}{\|u - u_h\|_{1,\mathcal{T}}}\right) \Big/ \log\left(\frac{\mathrm{DOFS}_{2h}^{-1/2}}{\mathrm{DOFS}_h^{-1/2}}\right) .$$

Here u_h denotes the approximate solution on the current refinement level and u_{2h} the approximate solution on the previous refinement level. Analogously, we define $EOC_{h,0}$ for the experimental order of convergence of the $L^2(\Omega)$-error in the primal variable, $EOC_{h,\mathbf{V}}$ for the $L^2(\Omega)$-error in the dual variable $\mathbf{q} \in \mathbf{V}$, and EOC for the error $\|(\mathbf{q} - \mathbf{q}_h, u - u_h)\|_{\mathbf{V} \times W}$ in the norm of the product space $\mathbf{V} \times W$.

The above definition reflects the fact that the number of globally coupled degrees of freedom (DOFS) is proportional to h^{-2} not respecting boundary effects.

5.2   Poisson equation with smooth solution

As a first example, we consider the Poisson equation on $\Omega = (0,1)^2$ with homogeneous Dirichlet boundary conditions, where $a = Id$, $d = 0$, and the right-hand side function f is chosen such that $u(x,y) = \sin(\pi x)\sin(\pi y)$ is the value of the exact solution at $(x,y) \in \Omega$. As initial triangulation we choose the set of eight congruent right-angled isosceles, all sharing the point $(0.5, 0.5)$ as a common vertex.

Since $u \in C^\infty(\Omega)$, the convergence rate of the individual methods are determined by the a priori estimates of Chapter 3. Table 5.1 shows the experimental orders of convergence for different polynomial degrees k using the hybridized IPDG, CG and NC methods. As expected, all three different methods converge with the same speed.

In Table 5.2, we show the results for the hybridized LDG method for $\tau = \gamma(k+1)^0 h^i$, $i \in \{-1,0\}$. Again the approximate primal variable is super-convergent. We note that, if the penalty parameter is chosen as $\tau = \gamma(k+1)^0 h^0$, both the L^2-error in the primal, as well as the L^2-error in the dual variable, is super-convergent in accordance to the results of [21].

In Table 5.3, we display the results for the hybridized RT and BDM methods. In case of the lowest order RT method, we also show the behavior of the post-processed primal variable using the post-processing of [57] instead of the original approximation $u_h \in W_h$, see Remark 4.1.6. It is obvious that the piecewise constant approximation u_h can not converge in the broken H^1-norm to the primal variable of the primal mixed formulation at all, since $\|\nabla(u - u_h)\|_{0,\mathcal{T}} = \|\nabla u\|_{0,\mathcal{T}}$.

k	NDOF	$\|u - u_h\|_0$	EOC_0	$\|\nabla(u - u_h)\|_0$	EOC_1	$\|q - q_h\|_0$	EOC_V
		H-CG method					
1	1	1.11e-01		8.96e-01		8.89e-01	
	9	7.09e-02	0.56	8.02e-01	0.14	7.99e-01	0.13
	49	1.84e-02	1.82	4.09e-01	0.91	4.08e-01	0.91
	225	4.64e-03	1.93	2.05e-01	0.97	2.05e-01	0.96
	961	1.16e-03	1.97	1.03e-01	0.98	1.03e-01	0.98
	3969	2.91e-04	1.98	5.14e-02	1.00	5.14e-02	1.00
	16129	7.27e-05	1.99	2.57e-02	1.00	2.57e-02	1.00
5	33	1.33e-05		6.03e-04		6.03e-04	
	169	1.37e-06	2.82	7.88e-05	2.53	7.88e-05	2.53
	753	2.15e-08	5.61	2.48e-06	4.67	2.48e-06	4.67
	3169	3.36e-10	5.81	7.76e-08	4.84	7.76e-08	4.84
	12993	5.25e-12	5.91	2.43e-09	4.92	2.43e-09	4.92
		H-NC method					
1	8	7.11e-02		9.19e-01		9.16e-01	
	40	2.54e-02	1.28	6.40e-01	0.45	6.39e-01	0.45
	176	6.54e-03	1.83	3.24e-01	0.92	3.24e-01	0.92
	736	1.65e-03	1.93	1.62e-01	0.97	1.62e-01	0.97
	3008	4.12e-04	1.97	8.13e-02	0.98	8.13e-02	0.98
	12160	1.03e-04	1.98	4.06e-02	0.99	4.06e-02	0.99
	48896	2.58e-05	1.99	2.03e-02	1.00	2.03e-02	1.00
3	24	1.26e-03		2.48e-02		2.48e-02	
	120	2.78e-04	1.88	1.17e-02	0.93	1.17e-02	0.93
	528	1.72e-05	3.76	1.48e-03	2.79	1.48e-03	2.79
	2208	1.07e-06	3.88	1.85e-04	2.91	1.85e-04	2.91
	9024	6.70e-08	3.94	2.32e-05	2.95	2.32e-05	2.95
	36480	4.19e-09	3.97	2.90e-06	2.98	2.90e-06	2.98
		H-IPDG method					
1	16	9.67e-02		8.77e-01		8.25e-01	
	80	6.61e-02	0.47	7.85e-01	0.14	7.48e-01	0.12
	352	1.72e-02	1.82	4.00e-01	0.91	3.82e-01	0.91
	1472	4.34e-03	1.92	2.01e-01	0.96	1.92e-01	0.96
	6016	1.09e-03	1.96	1.01e-01	0.98	9.61e-02	0.98
	24320	2.72e-04	1.99	5.03e-02	1.00	4.81e-02	0.99
3	32	1.76e-03		3.27e-02		3.20e-02	
	160	2.91e-04	2.24	1.29e-02	1.16	1.27e-02	1.15
	704	1.77e-05	3.78	1.62e-03	2.80	1.59e-03	2.80
	2944	1.09e-06	3.90	2.02e-04	2.91	1.99e-04	2.91
	12032	6.81e-08	3.94	2.53e-05	2.95	2.49e-05	2.95
	48640	4.26e-09	3.97	3.16e-06	2.98	3.11e-06	2.98
5	48	1.32e-05		6.00e-04		5.96e-04	
	240	1.36e-06	2.82	7.86e-05	2.53	7.79e-05	2.53
	1056	2.14e-08	5.60	2.48e-06	4.67	2.45e-06	4.67
	4416	3.34e-10	5.82	7.75e-08	4.84	7.68e-08	4.84
	18048	3.03e-11	3.41	2.43e-09	4.92	2.41e-09	4.92

Table 5.1: Example 1: H-CG, H-NC, and H-IPDG methods

k	NDOF	$\|u - u_h\|_0$	EOC_0	$\|\nabla(u - u_h)\|_0$	EOC_1	$\|\mathbf{q} - \mathbf{q}_h\|_0$	$EOC_{\mathbf{V}}$
\multicolumn{8}{c}{H-LDG method, where $\tau = \gamma h_T^0$}							
1	16	2.11e-02		7.96e-01		2.80e-01	
	80	1.40e-02	0.51	6.60e-01	0.23	1.91e-01	0.48
	352	2.17e-03	2.52	3.33e-01	0.92	5.59e-02	1.66
	1472	3.68e-04	2.48	1.67e-01	0.96	1.53e-02	1.81
	6016	7.68e-05	2.23	8.34e-02	0.99	4.04e-03	1.89
	24320	1.85e-05	2.04	4.17e-02	0.99	1.04e-03	1.94
5	48	9.85e-06		9.53e-04		3.12e-04	
	240	6.59e-07	3.36	1.04e-04	2.75	9.84e-06	4.30
	1056	1.02e-08	5.63	3.39e-06	4.62	1.65e-07	5.52
	4416	1.59e-10	5.82	1.08e-07	4.82	2.67e-09	5.76
	18048	2.47e-12	5.92	3.40e-09	4.91	4.37e-11	5.84
7	64	5.45e-08		1.28e-05		1.91e-06	
	320	8.79e-10	5.13	3.81e-07	4.37	2.39e-08	5.44
	1408	3.43e-12	7.49	3.10e-09	6.49	9.85e-11	7.41
\multicolumn{8}{c}{H-LDG method, where $\tau = \gamma h_T^{-1}$}							
1	16	3.20e-02		7.86e-01		3.25e-01	
	80	2.77e-02	0.18	6.75e-01	0.19	3.39e-01	-0.05
	352	7.15e-03	1.83	3.43e-01	0.91	1.70e-01	0.93
	1472	1.80e-03	1.93	1.72e-01	0.96	8.48e-02	0.97
	6016	4.51e-04	1.97	8.61e-02	0.98	4.24e-02	0.98
	24320	1.13e-04	1.98	4.31e-02	0.99	2.12e-02	0.99
	97792	2.82e-05	2.00	2.15e-02	1.00	1.06e-02	1.00
	392192	7.06e-06	1.99	1.08e-02	0.99	5.30e-03	1.00
5	48	6.59e-06		8.01e-04		3.25e-04	
	240	7.33e-07	2.73	9.42e-05	2.66	2.12e-05	3.39
	1056	1.16e-08	5.60	2.97e-06	4.67	6.63e-07	4.68
	4416	1.82e-10	5.81	9.32e-08	4.84	2.07e-08	4.85
	18048	2.85e-12	5.91	2.91e-09	4.92	6.48e-10	4.92
7	64	3.22e-08		9.96e-06		1.97e-06	
	320	1.02e-09	4.29	3.44e-07	4.18	5.49e-08	4.45
	1408	4.00e-12	7.48	2.71e-09	6.54	4.30e-10	6.55

Table 5.2: Example 1: H-LDG method

k	NDOF	$\|u - u_h\|_0$	EOC_0	$\|\nabla(u - u_h)\|_0$	EOC_1	$\|\mathbf{q} - \mathbf{q}_h\|_0$	$EOC_{\mathbf{V}}$
				H-RT method			
0	8	1.17e-01		9.97e-01		9.97e-01	
	40	1.81e-02	2.32	5.07e-01	0.84	5.07e-01	0.84
	176	2.94e-03	2.45	2.52e-01	0.94	2.52e-01	0.94
	736	5.89e-04	2.25	1.26e-01	0.97	1.26e-01	0.97
	3008	1.37e-04	2.07	6.30e-02	0.98	6.30e-02	0.98
	12160	3.37e-05	2.01	3.15e-02	0.99	3.15e-02	0.99
	48896	8.38e-06	2.00	1.57e-02	1.00	1.57e-02	1.00
	196096	2.09e-06	2.00	7.87e-03	0.99	7.87e-03	0.99
1	16	3.31e-02		8.22e-01		2.67e-01	
	80	1.96e-02	0.65	6.18e-01	0.35	5.78e-02	1.90
	352	4.96e-03	1.85	3.16e-01	0.91	1.39e-02	1.92
	1472	1.24e-03	1.94	1.59e-01	0.96	3.44e-03	1.95
	6016	3.11e-04	1.96	7.96e-02	0.98	8.58e-04	1.97
	24320	7.78e-05	1.98	3.98e-02	0.99	2.14e-04	1.99
	97792	1.94e-05	2.00	1.99e-02	1.00	5.36e-05	1.99
	392192	4.86e-06	1.99	9.95e-03	1.00	1.34e-05	2.00
3	32	6.88e-04		2.13e-02		6.82e-03	
	160	1.92e-04	1.59	1.01e-02	0.93	3.31e-04	3.76
	704	1.21e-05	3.73	1.28e-03	2.79	2.07e-05	3.74
	2944	7.61e-07	3.87	1.61e-04	2.90	1.29e-06	3.88
	12032	4.76e-08	3.94	2.01e-05	2.96	8.07e-08	3.94
	48640	2.97e-09	3.97	2.51e-06	2.98	5.05e-09	3.97
	195584	1.87e-10	3.97	3.14e-07	2.99	3.30e-10	3.92
				H-BDM method			
3	32	2.27e-02		5.45e-01		1.88e-02	
	160	2.17e-03	2.92	1.10e-01	1.99	9.08e-04	3.77
	704	2.75e-04	2.79	2.79e-02	1.85	5.69e-05	3.74
	2944	3.45e-05	2.90	6.99e-03	1.93	3.56e-06	3.87
	12032	4.31e-06	2.95	1.75e-03	1.97	2.22e-07	3.94
	48640	5.39e-07	2.98	4.38e-04	1.98	1.39e-08	3.97
5	48	5.75e-04		2.84e-02		3.94e-04	
	240	1.36e-05	4.65	1.36e-03	3.78	4.59e-06	5.53
	1056	4.30e-07	4.66	8.60e-05	3.73	7.26e-08	5.60
	4416	1.35e-08	4.84	5.40e-06	3.87	1.14e-09	5.81
	18048	4.22e-10	4.92	3.38e-07	3.94	2.03e-11	5.72
7	64	7.52e-06		5.97e-04		3.15e-06	
	320	4.35e-08	6.40	6.99e-06	5.53	9.15e-09	7.26
	1408	3.44e-10	6.53	1.11e-07	5.59	3.61e-11	7.47
	5888	2.69e-12	6.78	1.73e-09	5.82	7.66e-12	2.17

Table 5.3: Example 1: Hybridized mixed methods

5.3   Poisson equation on L-shaped domain

As a second example, we consider the Poisson equation on an L-shaped domain $\Omega = (-1,1)^2 \setminus (0,1) \times (-1,0)$, with homogeneous Dirichlet boundary conditions, where $a = Id$, $d = 0$ and the right-hand side function $f = -r^{\frac{2}{3}}\sin(\frac{2}{3}\varphi)(\Theta''(r) + \frac{7}{3}\Theta'(r)/r)$ is chosen such that $u = r^{\frac{2}{3}}\sin(\frac{2}{3}\varphi)\Theta(r)$ is the exact solution, where

$$
\Theta(r) = \begin{cases} 1 & , \, r < 1/4 \\ -6\bar{r}^5 + 15\bar{r}^4 - 10\bar{r}^3 + 1 & , \, 0 \leq \bar{r} < 1 \\ 0 & , \, r \geq 3/4 \end{cases} \quad , \, \bar{r} := 2(r - \tfrac{1}{4}) \quad .
$$

It is well-known that the exact solution u is in $H^1(\Omega)$, but not in $H^2(\Omega)$ and $\mathbf{q} \in H(div;\Omega)$ admits a singularity at the origin. Thus, this test problem is well-suited for adaptive hybridized methods, based on a posteriori error estimators. For numerical computations, the refinement parameter θ in the marking strategy is chosen as $\theta = 0.7$. A comparison of different refinement strategies will be given for a later example. For the H-IPDG methods, the penalty parameter is chosen as $\tau := \gamma(k + 1)^{2.5}h_{\mathcal{T}}^{-1}$, where $\gamma := 50$ in order to obtain a well-defined method, while for the H-LDG methods, τ is chosen as $\tau := \gamma h_{\mathcal{T}}^{-1}$, where $\gamma := 10$.

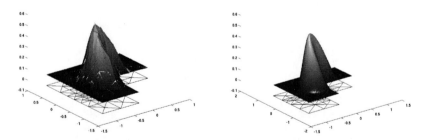

(a) u_h on fourth ref. level, using H-LDG-1   (b) u_h on fourth ref. level, using H-LDG-9

Figure 5.1: Example 2: L-shaped domain

Figure 5.1 shows the graph of the approximate solution u_h on the fourth refinement level for the H-LDG methods, using polynomials of degree $k = 1$ for both discrete spaces $\mathbf{V}_h$ and W_h, and polynomials of degree $k = 5$, respectively. Obviously the higher order approximation results in a better approximation to the exact solution, which is confirmed by tables 5.4, 5.5, and 5.6.

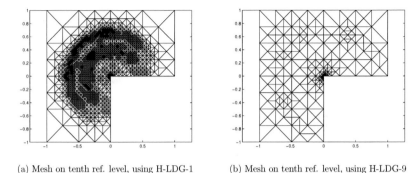

(a) Mesh on tenth ref. level, using H-LDG-1   (b) Mesh on tenth ref. level, using H-LDG-9

Figure 5.2: Example 2: L-shaped domain

Figure 5.2 shows the adaptively generated mesh on the tenth refinement level, using the same method as above, where $k = 1$ and $k = 10$, respectively. It can be seen that for the larger value of k, refinement is concentrated at the singularity at the origin, while for the lower order method the ring of a relatively strong varying gradient, caused by the polynomial cut-off-function Θ, can not be approximated well without strong refinement.

k	level	NDOF	$\|u - \tilde{u}_h\|_0$	$\|\mathbf{q} - \mathbf{q}_h\|_0$	$\|(\mathbf{q}, u) - (\mathbf{q}_h, u_h)\|$	η	EOC	EI
				H-RT method				
0	1	28	4.53e-01	2.19e+00	3.13e+00	8.52e+00		2.72
	2	73	6.70e-02	7.10e-01	1.01e+00	3.13e+00	2.37	3.11
	3	188	3.37e-02	4.50e-01	6.37e-01	1.98e+00	0.97	3.11
	11	29027	3.54e-04	3.00e-02	4.24e-02	1.36e-01	1.08	3.21
	12	57401	3.65e-04	2.19e-02	3.10e-02	9.88e-02	0.92	3.19
	13	104116	2.89e-04	1.60e-02	2.26e-02	7.27e-02	1.05	3.21
1	1	56	1.51e-01	1.01e+00	1.52e+00	2.73e+00		1.80
	2	138	8.17e-02	4.71e-01	8.37e-01	2.43e+00	1.32	2.90
	3	358	3.95e-02	2.04e-01	4.55e-01	1.62e+00	1.28	3.56
	11	44128	7.19e-05	2.54e-03	3.18e-02	1.20e-01	1.07	3.77
	12	89170	5.21e-05	1.56e-03	2.24e-02	8.57e-02	1.00	3.83
	13	159662	2.21e-05	9.71e-04	1.65e-02	6.25e-02	1.04	3.78
3	1	112	1.09e-02	1.29e-01	3.19e-01	1.79e+00		5.60
	2	268	1.66e-02	9.54e-02	1.51e-01	8.77e-01	1.72	5.80
	3	680	1.12e-03	2.93e-02	5.79e-02	5.96e-01	2.06	10.30
	10	37672	2.03e-05	1.41e-03	1.77e-03	5.69e-02	1.78	32.14
	11	63156	1.48e-05	9.42e-04	1.14e-03	4.27e-02	1.69	37.31
	12	110468	1.50e-05	5.97e-04	7.50e-04	3.15e-02	1.51	41.99
				H-BDM method				
3	1	112	1.80e-02	1.65e-01	4.29e-01	3.30e+00		7.69
	2	268	5.81e-03	5.14e-02	2.33e-01	1.01e+00	1.40	4.33
	3	604	2.26e-03	4.00e-02	1.29e-01	5.56e-01	1.46	4.31
	10	14412	3.22e-05	1.21e-03	6.94e-03	2.91e-02	2.09	4.19
	11	21040	1.76e-05	7.49e-04	4.56e-03	1.91e-02	2.22	4.19
	12	32184	1.01e-05	4.65e-04	2.94e-03	1.23e-02	2.07	4.18
5	1	168	5.76e-03	7.64e-02	2.13e-01	9.46e-01		4.44
	2	402	1.70e-03	2.64e-02	4.99e-02	2.34e-01	3.33	4.69
	3	846	8.97e-04	1.41e-02	3.24e-02	1.52e-01	1.16	4.69
	16	22284	1.84e-06	4.11e-05	1.29e-04	5.75e-04	4.52	4.46
	17	28842	1.27e-06	2.59e-05	8.34e-05	3.71e-04	3.38	4.45
	18	35004	6.12e-07	1.58e-05	5.26e-05	2.32e-04	4.76	4.41
9	1	280	3.31e-03	2.32e-02	3.89e-02	1.90e-01		4.88
	2	740	5.39e-04	1.24e-02	1.70e-02	8.38e-02	1.70	4.93
	3	1060	1.90e-04	7.30e-03	1.15e-02	6.20e-02	2.18	5.39
	17	22360	1.86e-06	2.35e-05	4.20e-05	2.17e-04	6.38	5.17
	18	26540	1.09e-06	1.52e-05	2.79e-05	1.43e-04	4.77	5.13
	19	31360	8.60e-07	9.96e-06	1.83e-05	9.11e-05	5.05	4.98

Table 5.4: Example 2: Hybridized mixed methods

k	level	NDOF	$\|u - \tilde{u}_h\|_0$	$\|\mathbf{q} - \mathbf{q}_h\|_0$	$\|(\mathbf{q}, u) - (\mathbf{q}_h, u_h)\|$	η	EOC	EI
— H-LDG method, $\tau = \gamma h_T^{-1}$								
1	1	56	9.98e-02	8.17e-01	1.47e+00	7.81e+00		5.31
	2	122	3.23e-02	3.79e-01	7.51e-01	4.77e+00	1.73	6.35
	3	336	1.90e-02	2.82e-01	5.73e-01	2.74e+00	0.53	4.78
	4	638	1.49e-02	1.96e-01	4.04e-01	1.91e+00	1.09	4.73
	11	45698	1.05e-04	1.63e-02	3.76e-02	1.81e-01	1.05	4.81
	12	85248	6.45e-05	1.18e-02	2.74e-02	1.31e-01	1.02	4.78
	13	157344	3.05e-05	8.63e-03	2.00e-02	9.60e-02	1.03	4.80
5	1	168	3.39e-03	6.74e-02	1.43e-01	8.26e-01		5.78
	2	402	1.69e-03	2.30e-02	4.01e-02	2.15e-01	2.91	5.36
	3	768	1.11e-03	1.40e-02	2.66e-02	1.48e-01	1.27	5.56
	4	1356	2.52e-04	9.20e-03	1.65e-02	9.23e-02	1.68	5.59
	23	59112	2.05e-07	2.89e-06	6.29e-06	3.18e-05	4.52	5.06
	24	71508	9.90e-08	1.93e-06	4.26e-06	2.11e-05	4.09	4.95
	25	87972	8.18e-08	1.28e-06	2.78e-06	1.35e-05	4.12	4.86
9	1	280	3.29e-03	2.30e-02	4.41e-02	2.11e-01		4.78
	2	670	6.04e-04	1.29e-02	1.78e-02	8.63e-02	2.08	4.85
	3	1050	1.89e-04	7.43e-03	1.14e-02	6.15e-02	1.98	5.39
	18	29530	8.89e-07	1.15e-05	2.38e-05	1.22e-04	5.90	5.13
	19	36380	3.69e-07	7.80e-06	1.63e-05	8.24e-05	3.63	5.06
	20	42760	4.21e-07	5.36e-06	1.13e-05	5.61e-05	4.53	4.96
— H-IPDG method, $\tau = \gamma h_T^{-1}$								
1	1	56	1.69e-01	1.18e+00	1.70e+00	9.39e+00		5.52
	2	122	6.30e-02	6.77e-01	9.79e-01	6.47e+00	1.42	6.61
	3	334	3.16e-02	4.85e-01	7.01e-01	4.11e+00	0.66	5.86
	8	9264	1.23e-03	8.48e-02	1.23e-01	7.97e-01	1.17	6.48
	9	17704	5.57e-04	6.04e-02	8.74e-02	5.73e-01	1.06	6.56
	10	31716	3.26e-04	4.47e-02	6.47e-02	4.24e-01	1.03	6.55
5	1	168	4.06e-03	8.89e-02	1.26e-01	1.17e+00		9.29
	2	366	4.85e-03	5.10e-02	7.27e-02	6.47e-01	1.41	8.90
	3	606	2.00e-03	2.74e-02	3.93e-02	4.24e-01	2.44	10.79
	17	14982	1.04e-05	1.20e-04	1.70e-04	1.58e-03	4.87	9.29
	18	18300	2.55e-06	7.30e-05	1.04e-04	1.03e-03	4.91	9.90
	19	21546	2.48e-06	4.86e-05	6.90e-05	6.81e-04	5.03	9.87
9	1	280	3.07e-03	2.52e-02	3.59e-02	4.27e-01		11.89
	2	550	1.35e-02	1.47e-02	2.10e-02	2.81e-01	1.59	13.38
	3	960	9.61e-04	1.12e-02	1.59e-02	2.00e-01	1.00	12.58
	15	14550	4.43e-06	9.76e-05	1.38e-04	1.41e-03	3.07	10.22
	16	16860	2.59e-06	6.77e-05	9.59e-05	9.67e-04	4.94	10.08
	17	19430	1.99e-06	4.36e-05	6.17e-05	5.92e-04	6.22	9.59

Table 5.5: Example 2: DG methods

k	level	NDOF	$\|u - \tilde{u}_h\|_0$	$\|\mathbf{q} - \mathbf{q}_h\|_0$	$\|(\mathbf{q}, u) - (\mathbf{q}_h, u_h)\|$	η	EOC	effind
\multicolumn{9}{c}{H-CG method}
1	1	5	1.64e-01	1.23e+00	1.74e+00	9.44e+00		5.43
	2	16	6.39e-02	7.12e-01	1.01e+00	6.75e+00	1.40	6.68
	3	50	3.19e-02	5.10e-01	7.22e-01	4.30e+00	0.67	5.96
	9	3215	5.89e-04	6.22e-02	8.79e-02	5.84e-01	0.97	6.64
	10	5713	3.13e-04	4.57e-02	6.47e-02	4.30e-01	1.07	6.65
	11	11193	1.58e-04	3.34e-02	4.73e-02	3.14e-01	0.93	6.64
5	1	117	4.33e-03	8.44e-02	1.19e-01	1.19e+00		10.00
	2	260	4.26e-03	4.58e-02	6.48e-02	6.38e-01	1.56	9.85
	3	501	1.06e-03	2.24e-02	3.17e-02	4.01e-01	2.20	12.65
	18	12446	4.34e-06	7.66e-05	1.08e-04	1.08e-03	5.02	10.00
	19	14918	3.58e-06	5.22e-05	7.39e-05	7.22e-04	4.19	9.77
	20	17492	1.95e-06	3.14e-05	4.45e-05	4.47e-04	6.38	10.04
9	1	252	1.52e-03	2.19e-02	3.11e-02	4.30e-01		13.83
	2	495	8.44e-04	1.40e-02	1.98e-02	2.89e-01	1.34	14.60
	3	864	1.56e-03	1.29e-02	1.83e-02	2.12e-01	0.28	11.58
	17	14877	4.89e-06	6.46e-05	9.14e-05	9.59e-04	6.01	10.49
	18	17109	5.40e-06	4.78e-05	6.78e-05	6.38e-04	4.27	9.41
	19	20511	3.94e-06	3.40e-05	4.83e-05	4.34e-04	3.74	8.99
\multicolumn{9}{c}{H-NC method}
1	1	28	7.79e-02	1.07e+00	1.52e+00	9.92e+00		6.53
	2	67	5.81e-02	7.41e-01	1.05e+00	6.94e+00	0.85	6.61
	3	182	3.61e-02	4.86e-01	6.88e-01	4.38e+00	0.85	6.37
	10	12324	4.53e-04	5.33e-02	7.54e-02	4.69e-01	1.06	6.22
	11	20791	2.73e-04	4.09e-02	5.78e-02	3.48e-01	1.02	6.02
	12	39685	1.36e-04	2.93e-02	4.15e-02	2.58e-01	1.02	6.22
5	1	140	4.01e-03	9.03e-02	1.28e-01	1.23e+00		9.61
	2	345	2.03e-03	4.76e-02	6.74e-02	6.55e-01	1.42	9.72
	3	575	1.97e-03	2.82e-02	3.99e-02	4.27e-01	2.05	10.70
	21	23620	1.50e-06	2.21e-05	3.13e-05	2.93e-04	4.93	9.36
	22	29225	6.75e-07	1.41e-05	1.99e-05	1.90e-04	4.25	9.55
	23	34650	7.33e-07	9.43e-06	1.34e-05	1.23e-04	4.65	9.18
9	1	252	3.20e-03	2.59e-02	3.67e-02	4.90e-01		13.35
	2	540	2.36e-03	1.86e-02	2.64e-02	3.47e-01	0.86	13.14
	3	990	1.70e-03	1.36e-02	1.94e-02	2.27e-01	1.02	11.70
	17	16632	1.42e-06	5.38e-05	7.61e-05	7.24e-04	4.74	9.51
	18	19143	2.72e-06	3.48e-05	4.93e-05	4.74e-04	6.17	9.61
	19	22752	1.04e-06	2.09e-05	2.96e-05	3.07e-04	5.91	10.37

Table 5.6: Example 2: CG and NC methods

5.4 Interior layer

Consider the reaction-diffusion equation (2.1) as a third example, where $a = Id$, $d = 2$ and Ω is a quadrilateral, shaped as depicted in Figure 5.3. Here f and the boundary data g are chosen such that the exact solution u satisfies

$$u(x,y) = \arctan\left(20(x^2 + y^2) - 1\right) \quad , \ \forall(x,y) \in \Omega \ .$$

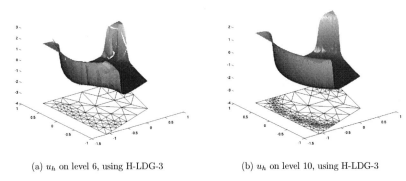

(a) u_h on level 6, using H-LDG-3   (b) u_h on level 10, using H-LDG-3

Figure 5.3: Example 3: interior layer

For all numerical tests on this test problem, the initial triangulation $\mathcal{T}$ is generated by splitting Ω into two triangles, using the short diagonal and bisecting both elements, as well as their children. In Figure 5.3, we display the approximate solution u_h on different refinement levels, using the same adaptive H-LDG method as above, where $\tau = \gamma h_{\mathcal{T}}^{-1}$ and $k = 3$. Due to strong data oscillation, the exact solution is only poorly approximated on the lower refinement level.

For the H-RT method, we see (cf. Figure 5.4a) that the post-processed lowest order approximation converges as fast as the standard H-RT-1 method. From 5.4b we see that the efficiency index EI stays bounded during adaptive refinement, for each polynomial degree k. We note that for the lowest order post-processed $H^1(\Omega) \times H(div; \Omega)$-conforming approximation $(\tilde{u}_h, \mathbf{q}_h)$, the estimator η of Remark 4.1.6 is reliable without any hidden constants.

In Figure 5.5, we show the behavior of the error estimator η in dependence of the number of globally coupled degrees of freedom (DOFs= $dim(M_h)$) for two different refinement parameters θ in comparison to global refinement, using the post-processed H-RT-0 method

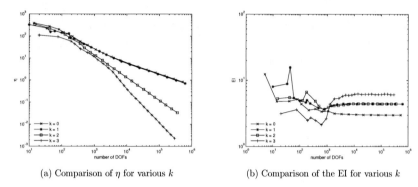

(a) Comparison of η for various k (b) Comparison of the EI for various k

Figure 5.4: Example 3: H-RT

and the H-RT-3 method, respectively. For $\theta = 0.3$ only a small amount of elements is
refined in each step. Thus, only few DOFs are needed to reach a certain value of η,
with the drawback of reaching the maximum number of loops in the adaptive cycle early.
For this test problem, where the exact solution admits high regularity, the experimental
convergence rates using global refinement, are not worse than the convergence rates for
the adaptive methods, after an initial phase of mesh adaption. For practical calculations,
however, a prescribed tolerance TOL$\leq \eta$ can be met with significantly less DOFs, using the
adaptive method. For $k = 3$, this effect is even stronger than for lower k.

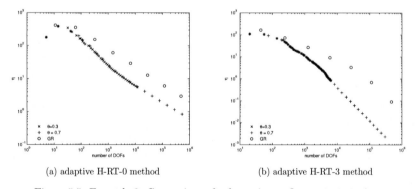

(a) adaptive H-RT-0 method   (b) adaptive H-RT-3 method

Figure 5.5: Example 3: Comparison of η for various refinement strategies

Figure 5.6 displays the convergence rates for the estimator η in terms of globally coupled degrees of freedom for the H-IPDG methods and the H-LDG methods, respectively, for different values of k. For the H-IPDG methods, the penalty parameter is chosen according to $\tau = \gamma(k + 1)^{2.5}h_T^{-1}$, where $\gamma = 50$, while for the H-LDG method $\tau = \gamma h_T^{-1}$, where $\gamma = 10$. The refinement parameter of the marking strategy is set to $\theta = 0.7$. We observe in both cases that the convergence rates for η grow with increasing k. We observe linear convergence for $k = 1$. For larger k the methods converge increasingly super-linear. Similar behavior can be observed for the other methods under consideration, see tables 5.7, 5.8, and 5.9.

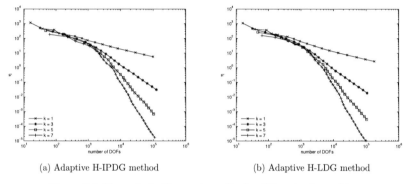

(a) Adaptive H-IPDG method   (b) Adaptive H-LDG method

Figure 5.6: Example 3: Comparison of η for different values of k.

In Theorem 4.4.1 it has been shown that the error estimator η is efficient up to a constant, which depends on the polynomial degree k. In Figure 5.7, it can be seen that the efficiency index EI stays bounded for each k. This confirms the result of Theorem 4.4.1. Furthermore, the efficiency index for the H-IPDG method increases with growing polynomial degree k, not respecting irregularities on relatively coarse meshes, which are due to data oscillation. This is in accordance with the theoretical result in Chapter 4, where it has been shown that the error estimator can be bounded by the actual error only up to constant, which grows with increasing polynomial degree k. For the H-LDG method, we observe that the efficiency index does not depend on the polynomial degree k. This remarkable property of the H-LDG method can be observed for all test problems in this chapter.

k	level	NDOF	$\|u - \tilde{u}_h\|_0$	$\|\mathbf{q} - \mathbf{q}_h\|_0$	$\|(\mathbf{q}, u) - (\mathbf{q}_h, u_h)\|$	η	EOC	EI
				H-RT method				
0	1	5	1.10e+00	1.03e+01	1.45e+01	1.80e+02	0.00	12.38
	2	14	1.33e+01	5.67e+01	8.08e+01	3.88e+02	-3.33	4.80
	3	41	5.71e+00	3.75e+01	5.70e+01	2.76e+02	0.65	4.84
	4	82	2.52e+00	2.49e+01	3.88e+01	2.01e+02	1.11	5.19
	17	148517	1.98e-03	2.35e-01	5.25e-01	1.58e+00	1.00	3.01
	18	278869	9.39e-04	1.71e-01	3.83e-01	1.15e+00	1.00	3.00
	19	533299	5.64e-04	1.24e-01	2.78e-01	8.36e-01	0.99	3.01
1	1	10	4.34e+00	3.75e+01	4.12e+01	3.31e+02	0.00	8.03
	2	34	2.99e+00	2.36e+01	2.57e+01	2.28e+02	0.77	8.86
	3	44	1.02e+00	6.58e+00	9.98e+00	1.57e+02	7.35	15.74
	4	60	3.67e+00	2.38e+01	3.12e+01	1.70e+02	-7.35	5.45
	19	173760	4.47e-04	1.10e-02	3.08e-01	1.36e+00	1.01	4.41
	20	328098	2.21e-04	6.43e-03	2.23e-01	9.84e-01	1.02	4.41
	21	622190	1.30e-04	3.92e-03	1.62e-01	7.14e-01	1.00	4.41
3	1	20	4.16e+00	2.56e+01	3.50e+01	1.10e+02	0.00	3.14
	2	76	1.12e+00	1.68e+01	2.58e+01	9.31e+01	0.46	3.61
	3	192	1.05e+00	1.56e+01	2.37e+01	5.90e+01	0.18	2.49
	4	252	7.70e-01	1.09e+01	1.62e+01	4.44e+01	2.78	2.74
	25	171868	1.14e-06	4.41e-05	8.37e-04	5.19e-03	3.07	6.20
	26	226900	7.32e-07	2.71e-05	5.64e-04	3.43e-03	2.85	6.09
	27	297520	3.65e-07	1.75e-05	3.80e-04	2.32e-03	2.90	6.10
				H-BDM method				
3	1	32	3.76e+00	2.63e+01	3.68e+01	5.83e+02	0.00	15.84
	2	80	1.63e+00	1.67e+01	2.34e+01	3.27e+02	0.99	13.97
	3	164	1.47e+00	1.32e+01	1.97e+01	2.06e+02	0.48	10.46
	4	244	1.02e+00	1.04e+01	1.57e+01	1.48e+02	1.14	9.43
	15	18224	6.70e-04	7.47e-03	2.20e-01	1.37e+00	2.11	6.23
	16	28788	2.82e-04	3.22e-03	1.34e-01	8.28e-01	2.17	6.18
	17	47236	1.72e-04	1.52e-03	8.08e-02	5.03e-01	2.04	6.23
5	1	48	2.28e+00	2.64e+01	4.06e+01	2.58e+02	0.00	6.35
	2	174	1.41e+00	1.70e+01	2.55e+01	1.51e+02	0.72	5.92
	3	252	5.76e-01	9.55e+00	1.48e+01	1.07e+02	2.94	7.23
	4	402	3.70e-01	6.29e+00	1.01e+01	7.10e+01	1.64	7.03
	18	26460	1.10e-05	2.67e-04	6.84e-03	4.30e-02	4.77	6.29
	19	36576	6.10e-06	1.16e-04	3.86e-03	2.43e-02	3.53	6.30
	20	48636	2.84e-06	5.28e-05	2.13e-03	1.34e-02	4.17	6.29
9	1	80	1.48e+00	1.47e+01	2.17e+01	1.29e+02	0.00	5.94
	2	200	5.20e-01	7.95e+00	1.18e+01	7.16e+01	1.33	6.07
	3	520	2.30e-01	3.51e+00	5.55e+00	3.84e+01	1.58	6.92
	4	760	1.09e-01	1.94e+00	3.37e+00	2.72e+01	2.63	8.07
	24	29530	3.73e-08	2.96e-06	5.25e-05	3.39e-04	9.86	6.46
	25	33540	3.10e-08	1.03e-06	3.02e-05	1.95e-04	8.69	6.46
	26	39340	1.31e-08	6.55e-07	1.69e-05	1.10e-04	7.28	6.51

Table 5.7: Example 3: mixed methods

k	level	NDOF	$\|u - \tilde{u}_h\|_0$	$\|\mathbf{q} - \mathbf{q}_h\|_0$	$\|(\mathbf{q},u)-(\mathbf{q}_h,u_h)\|$	η	EOC	EI
\multicolumn{9}{c}{H-LDG method, $\tau = \gamma h_T^{-1}$}								
1	1	16	6.83e+00	2.46e+01	4.33e+01	1.10e+03	0.00	25.40
	2	38	2.62e+00	1.75e+01	2.45e+01	4.35e+02	1.32	17.76
	3	88	2.59e+00	1.70e+01	2.55e+01	3.73e+02	-0.10	14.63
	4	126	1.32e+00	1.40e+01	2.06e+01	2.14e+02	1.19	10.39
	15	51150	1.37e-03	3.08e-01	6.81e-01	5.39e+00	1.01	7.91
	16	95878	7.67e-04	2.22e-01	4.94e-01	3.87e+00	1.02	7.83
	17	180568	4.24e-04	1.61e-01	3.58e-01	2.80e+00	1.02	7.82
5	1	48	2.39e+00	2.72e+01	4.81e+01	2.68e+02	0.00	5.57
	2	174	1.43e+00	1.76e+01	2.99e+01	1.59e+02	0.74	5.32
	3	252	5.78e-01	9.84e+00	1.77e+01	1.12e+02	2.83	6.33
	4	402	3.68e-01	6.37e+00	1.14e+01	7.34e+01	1.88	6.44
	26	72330	9.62e-08	2.38e-05	1.02e-04	8.02e-04	5.36	7.86
	27	88632	6.25e-08	1.44e-05	6.22e-05	4.90e-04	4.87	7.88
	28	103968	3.38e-08	9.00e-06	4.00e-05	3.12e-04	5.53	7.80
9	1	80	1.49e+00	1.48e+01	2.79e+01	1.70e+02	0.00	6.09
	2	200	5.17e-01	7.99e+00	1.44e+01	9.19e+01	1.44	6.38
	3	520	2.30e-01	3.55e+00	7.39e+00	5.11e+01	1.40	6.91
	4	760	1.08e-01	1.95e+00	4.41e+00	3.40e+01	2.72	7.71
	27	34910	4.16e-09	9.49e-07	5.79e-06	4.41e-05	9.21	7.62
	28	39000	2.20e-09	5.39e-07	3.51e-06	2.62e-05	9.04	7.46
	29	44410	1.03e-09	3.01e-07	2.02e-06	1.53e-05	8.51	7.57
\multicolumn{9}{c}{H-IPDG method, $\tau = \gamma h_T^{-1}$}								
1	1	16	3.89e+00	1.35e+01	2.17e+01	1.17e+03	0.00	53.92
	2	38	2.16e+00	1.44e+01	2.01e+01	4.56e+02	0.18	22.69
	3	88	2.27e+00	1.53e+01	2.20e+01	3.83e+02	-0.22	17.41
	4	126	1.20e+00	1.35e+01	1.93e+01	2.26e+02	0.73	11.71
	14	28646	3.71e-03	6.98e-01	1.12e+00	1.09e+01	0.97	9.73
	15	53158	2.27e-03	5.03e-01	8.06e-01	7.91e+00	1.06	9.81
	16	99792	1.25e-03	3.66e-01	5.87e-01	5.73e+00	1.01	9.76
5	1	48	2.24e+00	2.66e+01	3.79e+01	3.48e+02	0.00	9.18
	2	174	1.38e+00	1.73e+01	2.46e+01	1.92e+02	0.67	7.80
	3	252	5.68e-01	9.78e+00	1.39e+01	1.43e+02	3.08	10.29
	4	402	3.67e-01	6.49e+00	9.22e+00	8.70e+01	1.76	9.44
	26	72486	1.83e-07	7.77e-05	1.13e-04	1.81e-03	5.64	16.02
	27	87900	1.20e-07	4.83e-05	7.00e-05	1.12e-03	4.97	16.00
	28	103170	5.87e-08	3.09e-05	4.47e-05	7.11e-04	5.60	15.91
9	1	80	1.48e+00	1.48e+01	2.10e+01	1.75e+02	0.00	8.33
	2	200	5.17e-01	8.00e+00	1.13e+01	9.27e+01	1.35	8.20
	3	520	2.30e-01	3.72e+00	5.27e+00	5.09e+01	1.60	9.66
	4	760	1.09e-01	2.07e+00	2.94e+00	3.48e+01	3.08	11.84
	27	29510	1.84e-08	7.08e-06	1.01e-05	2.26e-04	11.24	22.38
	28	32390	1.09e-08	4.47e-06	6.39e-06	1.39e-04	9.83	21.75
	29	35620	7.87e-09	3.08e-06	4.41e-06	9.22e-05	7.80	20.91
	30	39880	4.19e-09	1.95e-06	2.78e-06	5.91e-05	8.17	21.26

Table 5.8: Example 3: DG methods

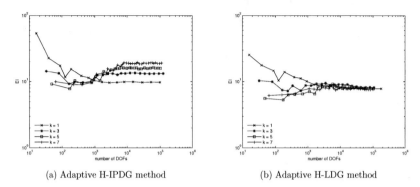

(a) Adaptive H-IPDG method   (b) Adaptive H-LDG method

Figure 5.7: Example 3: Comparison of the efficiency index for different values of k.

5.5   Discontinuous diffusion tensor

As a last example, we consider the case of a discontinuous diffusion tensor a in equation (2.1), a test problem taken from [57]. Therefore, the domain $\Omega = (-1, 1)^2$ is divided into four subdomains Ω_i, $1 \leq i \leq 4$, corresponding to the four quadrants of the Cartesian coordinate system, ordered counterclockwise, beginning from the quadrant with positive values in both coordinates. We set $a = a_i Id$ on Ω_i, where $a_1 = a_3 = 100$ and $a_2 = a_4 = 1$. Furthermore, we choose $d = 0$ and $f = 0$, while the boundary data g is chosen such that

$$u(r, \phi) = r^\alpha \big(b_i \sin(\alpha\phi) + c_i \cos(\alpha\phi)\big) \quad \text{on } \Omega_i, \ 1 \leq i \leq 4 ,$$

is the exact solution in polar coordinates, where $\alpha = 0.12690207$. The coefficients b_i and c_i are chosen according to the following table:

i	1	2	3	4
b_i	0.1	-9.60396040	-0.48035487	7.70156488
c_i	1	2.96039604	-0.88275659	-6.45646175

Thus, $u \in H^1(\Omega)$, while $\mathbf{q} \in H(div; \Omega)$ admits a singularity at the origin. The initial triangulation $\mathcal{T}$ of Ω consists of eight congruent right-angled isosceles, all sharing the point 0 as a common vertex. Figure 5.8 displays a visualization of the numerical solution computed after globally bisecting the triangulation twice, using the H-LDG method of order $k = 3$ with less than 700 degrees of freedom. Obviously, the method is not capable of resolving the strong variation of u around the origin.

k	level	NDOF	$\|u - \tilde{u}_h\|_0$	$\|\mathbf{q} - \mathbf{q}_h\|_0$	$\|(\mathbf{q}, u) - (\mathbf{q}_h, u_h)\|$	η	EOC	EI
			H-NC method					
1	1	8	1.64e+01	6.39e+01	9.19e+01	1.34e+03	0.00	14.58
	2	19	5.20e+00	3.14e+01	4.48e+01	5.28e+02	1.66	11.79
	3	52	1.67e+00	1.56e+01	2.22e+01	3.82e+02	1.39	17.21
	4	72	1.46e+00	1.51e+01	2.14e+01	2.49e+02	0.23	11.64
	14	12607	4.45e-03	9.35e-01	1.32e+00	1.37e+01	1.02	10.38
	15	23263	2.66e-03	6.85e-01	9.69e-01	9.98e+00	1.01	10.30
	16	43374	1.56e-03	5.00e-01	7.07e-01	7.29e+00	1.01	10.31
5	1	40	2.19e+00	2.70e+01	3.82e+01	3.55e+02	0.00	9.29
	2	145	1.32e+00	1.76e+01	2.49e+01	1.96e+02	0.66	7.87
	3	210	5.77e-01	9.81e+00	1.39e+01	1.44e+02	3.15	10.36
	4	335	3.68e-01	6.55e+00	9.27e+00	8.98e+01	1.73	9.69
	21	27240	2.09e-06	5.69e-04	8.05e-04	1.31e-02	5.31	16.27
	22	33185	9.39e-07	3.39e-04	4.80e-04	7.89e-03	5.24	16.44
	23	41540	5.10e-07	2.13e-04	3.01e-04	5.02e-03	4.16	16.68
7	1	56	1.08e+00	1.35e+01	1.91e+01	1.79e+02	0.00	9.37
	2	140	8.91e-01	1.08e+01	1.53e+01	1.29e+02	0.48	8.43
	3	364	5.28e-01	6.05e+00	8.57e+00	7.38e+01	1.21	8.61
	4	602	9.25e-02	2.72e+00	3.85e+00	4.67e+01	3.18	12.13
	25	32788	6.28e-08	2.64e-05	3.73e-05	7.14e-04	6.49	19.14
	26	38787	4.19e-08	1.52e-05	2.15e-05	4.20e-04	6.56	19.53
	27	44912	1.65e-08	9.14e-06	1.29e-05	2.56e-04	6.97	19.84
9	1	72	1.49e+00	1.49e+01	2.11e+01	1.70e+02	0.00	8.06
	2	180	5.29e-01	8.08e+00	1.14e+01	9.17e+01	1.34	8.04
	3	468	2.30e-01	3.74e+00	5.30e+00	5.22e+01	1.60	9.85
	4	684	1.09e-01	2.09e+00	2.96e+00	3.54e+01	3.07	11.96
	29	34290	4.66e-09	2.37e-06	3.35e-06	7.23e-05	9.75	21.58
	30	38772	2.92e-09	1.43e-06	2.02e-06	4.46e-05	8.24	22.08
	31	42048	2.44e-09	9.10e-07	1.29e-06	2.87e-05	11.06	22.25

Table 5.9: Example 3: NC method

For the visualization of the approximate flux $\mathbf{q}_h$, each arrow, scaled by the element diameter h_T, indicates the value of $\mathbf{q}_h$ at that point. The direction of each arrow indicates the direction of $\mathbf{q}_h$, while the color represents the Euclidean norm of $\mathbf{q}_h$ at that point, respectively.

In comparison to the approximation depicted in Figure 5.8, we display a visualization of the numerical solution to P7, for the adaptive H-LDG-3 method, again using less than 700 degrees of freedom. Seemingly the adaptive method provides a better approximation. Refinement is concentrated in the vicinity of the singularity at the origin. In Figure 5.10, we show the performance of the error estimator η in terms of globally coupled degrees of freedom, using different refinement strategies for the H-LDG-7 method and for the H-RT-0 method, respectively. Obviously, the adaptive methods yield much more accurate

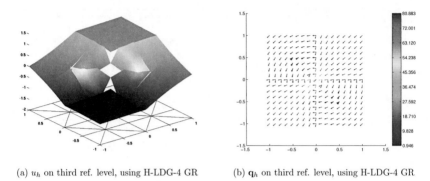

(a) u_h on third ref. level, using H-LDG-4 GR   (b) $\mathbf{q}_h$ on third ref. level, using H-LDG-4 GR

Figure 5.8: Example 4: discontinuous diffusion tensor a

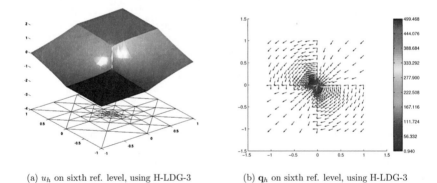

(a) u_h on sixth ref. level, using H-LDG-3   (b) $\mathbf{q}_h$ on sixth ref. level, using H-LDG-3

Figure 5.9: Example 4: Discontinuous diffusion tensor a

solutions.

Figure 5.11 shows the performance of the error estimator in terms of the globally coupled degrees of freedom for different polynomial degrees k, where the refinement parameter $\theta = 0.7$, using the H-BDM methods and the H-RT methods, respectively. For low k, the convergence rate stabilizes after an initial phase of mesh adaption during the adaptive cycle. The final rate grows with increasing polynomial degree. Seemingly the final convergence rate has not been reached for the adaptive H-BDM-5 and the H-BDM-7 method after the maximal number of iterations. Thus the former method seems to converge faster than the latter method.

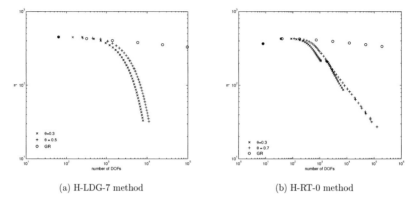

(a) H-LDG-7 method   (b) H-RT-0 method

Figure 5.10: Example 4: Convergence rates for various refinement strategies

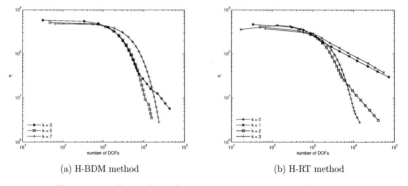

(a) H-BDM method   (b) H-RT method

Figure 5.11: Example 4: Comparison of η for various k, $\theta = 0.7$

Figure 5.12 displays the efficiency indices in terms of the globally coupled degrees of freedom for different polynomial degrees k, using the H-LDG methods, where $\tau = \gamma h_{\mathcal{T}}^{-1}$, and the H-RT methods, respectively. The efficiency indices remain bounded for any value of k. This observation confirms the theoretical expectation in case of the H-LDG method. Again, for the H-LDG method, the efficiency index does not grow with increasing polynomial degree k. For the H-RT methods, the efficiency indices remain bounded as well. For the post-processed H-RT-0 method, the efficiency index, not containing any hidden constants, is close to 1. For a quantitative comparison of the numerical test-results we refer to tables

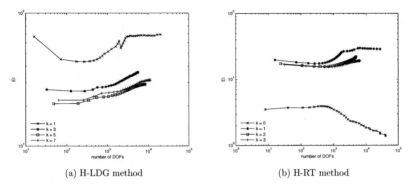

(a) H-LDG method   (b) H-RT method

Figure 5.12: Example 4: Comparison of the EI for various k, $\theta = 0.7$

5.10, 5.11, and 5.12.

k	level	NDOF	$\|u - \tilde{u}_h\|_0$	$\|\mathbf{q} - \mathbf{q}_h\|_0$	$\|(\mathbf{q}, u) - (\mathbf{q}_h, u_h)\|$	η	EOC	effind
			H-RT methods					
0	1	8	2.59e-01	1.35e+02	1.35e+02	4.76e+02	0.00	3.53
	2	40	1.44e-01	1.33e+02	1.33e+02	4.98e+02	0.02	3.74
	3	112	9.93e-02	1.26e+02	1.26e+02	4.72e+02	0.10	3.75
	39	49342	4.26e-04	3.67e+01	3.67e+01	4.26e+01	0.46	1.16
	40	49504	3.94e-04	3.67e+01	3.67e+01	4.09e+01	0.00	1.11
	41	55772	4.07e-04	3.53e+01	3.53e+01	3.79e+01	0.65	1.07
	42	89142	2.79e-04	3.14e+01	3.14e+01	3.23e+01	0.50	1.03
1	1	16	1.43e-01	2.47e+01	2.48e+01	4.91e+02	0.00	19.83
	2	72	9.91e-02	2.49e+01	2.49e+01	4.56e+02	-0.01	18.28
	3	168	7.84e-02	2.44e+01	2.44e+01	4.32e+02	0.05	17.67
	40	52524	3.82e-04	1.08e+00	1.09e+00	3.40e+01	2.92	31.15
	41	73800	2.71e-04	9.69e-01	9.79e-01	2.99e+01	0.64	30.54
	42	77844	2.57e-04	8.89e-01	8.99e-01	2.84e+01	3.20	31.60
	43	86960	2.18e-04	8.16e-01	8.25e-01	2.65e+01	1.54	32.11
3	1	32	8.07e-02	2.28e+01	2.28e+01	3.83e+02	0.00	16.79
	2	144	6.83e-02	2.23e+01	2.23e+01	3.56e+02	0.03	15.96
	3	336	5.86e-02	2.14e+01	2.14e+01	3.34e+02	0.10	15.60
	62	12312	1.80e-05	1.73e-01	1.74e-01	3.59e+00	3.27	20.67
	63	12960	1.45e-05	1.59e-01	1.60e-01	3.29e+00	3.29	20.62
	64	13608	1.36e-05	1.46e-01	1.47e-01	3.02e+00	3.50	20.61
	65	14256	1.34e-05	1.34e-01	1.35e-01	2.78e+00	3.69	20.67
			H-BDM methods					
3	1	32	8.52e-02	2.19e+01	2.19e+01	5.81e+02	0.00	26.53
	2	128	7.15e-02	2.17e+01	2.17e+01	5.66e+02	0.01	26.08
	3	228	6.28e-02	2.06e+01	2.06e+01	5.65e+02	0.18	27.43
	61	33848	6.40e-05	1.36e-01	1.39e-01	7.45e+00	2.31	53.60
	62	36624	5.81e-05	1.25e-01	1.27e-01	6.83e+00	2.29	53.78
	63	39764	4.13e-05	1.14e-01	1.16e-01	6.24e+00	2.20	53.79
	64	43148	3.41e-05	1.05e-01	1.06e-01	5.70e+00	2.21	53.77
5	1	48	6.83e-02	2.38e+01	2.38e+01	5.15e+02	0.00	21.64
	2	192	5.98e-02	2.29e+01	2.30e+01	5.02e+02	0.05	21.83
	3	414	5.21e-02	2.18e+01	2.18e+01	4.96e+02	0.14	22.75
	62	15084	2.71e-05	1.27e-01	1.28e-01	4.08e+00	9.79	31.88
	63	15336	2.70e-05	1.16e-01	1.17e-01	3.80e+00	10.85	32.48
	64	15588	2.70e-05	1.07e-01	1.07e-01	3.55e+00	10.96	33.18
	65	15840	2.70e-05	9.77e-02	9.83e-02	3.33e+00	10.58	33.88
7	1	64	6.04e-02	2.07e+01	2.07e+01	4.81e+02	0.00	23.24
	2	256	5.31e-02	2.00e+01	2.00e+01	4.68e+02	0.05	23.40
	3	600	4.58e-02	1.88e+01	1.88e+01	4.42e+02	0.15	23.51
	62	22992	1.87e-06	9.61e-02	9.65e-02	3.36e+00	10.02	34.82
	63	23376	1.59e-06	8.80e-02	8.84e-02	3.08e+00	10.59	34.84
	64	23760	1.37e-06	8.06e-02	8.09e-02	2.82e+00	10.88	34.86
	65	24144	1.18e-06	7.38e-02	7.41e-02	2.58e+00	10.95	34.82

Table 5.10: Example 4: mixed methods

k	level	NDOF	$\|u - \tilde{u}_h\|_0$	$\|\mathbf{q} - \mathbf{q}_h\|_0$	$\|(\mathbf{q},u)-(\mathbf{q}_h,u_h)\|$	η	EOC	effind
\multicolumn{9}{c}{H-LDG method, $\tau = \gamma h_T^{-1}$}							
1	1	16	8.44e-02	1.79e+01	1.80e+01	1.20e+03	0.00	66.67
	2	72	8.20e-02	1.87e+01	1.88e+01	8.46e+02	-0.06	45.00
	3	176	6.38e-02	1.85e+01	1.86e+01	8.07e+02	0.02	43.39
	4	272	5.39e-02	1.79e+01	1.79e+01	7.74e+02	0.18	43.24
	33	14280	7.52e-04	1.49e+00	1.52e+00	1.03e+02	1.22	67.76
	34	16872	6.67e-04	1.34e+00	1.37e+00	9.44e+01	1.25	68.91
	35	20326	5.83e-04	1.22e+00	1.25e+00	8.64e+01	0.98	69.12
5	1	48	6.74e-02	2.33e+01	2.33e+01	4.86e+02	0.00	20.86
	2	192	5.87e-02	2.24e+01	2.25e+01	4.71e+02	0.05	20.93
	3	342	5.22e-02	2.10e+01	2.11e+01	4.66e+02	0.22	22.09
	4	540	4.55e-02	1.94e+01	1.94e+01	4.51e+02	0.37	23.25
	33	7866	3.28e-04	1.43e+00	1.44e+00	4.18e+01	5.31	29.03
	34	8118	2.75e-04	1.31e+00	1.31e+00	3.83e+01	6.00	29.24
	35	8370	2.31e-04	1.20e+00	1.20e+00	3.50e+01	5.74	29.17
7	1	64	6.02e-02	2.04e+01	2.04e+01	4.52e+02	0.00	22.16
	2	256	5.27e-02	1.97e+01	1.97e+01	4.37e+02	0.05	22.18
	3	456	4.70e-02	1.84e+01	1.84e+01	4.35e+02	0.24	23.64
	4	720	4.09e-02	1.69e+01	1.69e+01	4.23e+02	0.37	25.03
	33	10488	2.90e-04	1.22e+00	1.23e+00	3.84e+01	5.26	31.22
	34	10824	2.43e-04	1.12e+00	1.12e+00	3.52e+01	5.94	31.43
	35	11160	2.04e-04	1.02e+00	1.03e+00	3.22e+01	5.48	31.26
\multicolumn{9}{c}{H-IPDG method, $\tau = \gamma h_T^{-1}$}							
1	1	16	1.38e-01	3.59e+01	3.60e+01	1.33e+04	0.00	369.44
	2	50	1.31e-01	3.33e+01	3.33e+01	1.19e+04	0.14	357.36
	3	72	1.08e-01	3.11e+01	3.11e+01	1.09e+04	0.37	350.48
	4	150	9.42e-02	2.90e+01	2.91e+01	9.49e+03	0.18	326.12
	23	11870	1.58e-02	5.02e+00	5.07e+00	2.04e+03	0.95	402.37
	24	14618	1.56e-02	4.58e+00	4.63e+00	1.87e+03	0.87	403.89
	25	17602	1.55e-02	4.18e+00	4.24e+00	1.71e+03	0.95	403.30
5	1	48	8.11e-02	3.00e+01	3.00e+01	1.81e+04	0.00	603.33
	2	192	6.71e-02	2.69e+01	2.69e+01	1.62e+04	0.16	602.23
	3	342	5.70e-02	2.44e+01	2.44e+01	1.44e+04	0.34	590.16
	4	558	4.71e-02	2.17e+01	2.17e+01	1.26e+04	0.48	580.65
	23	6048	1.44e-03	2.99e+00	3.00e+00	2.04e+03	3.91	680.00
	24	6336	1.20e-03	2.73e+00	2.74e+00	1.87e+03	3.90	682.48
	25	6624	1.01e-03	2.49e+00	2.50e+00	1.71e+03	4.12	684.00
7	1	64	7.10e-02	3.00e+01	3.00e+01	2.06e+04	0.00	686.67
	2	256	5.91e-02	2.69e+01	2.69e+01	1.86e+04	0.16	691.45
	3	456	5.04e-02	2.44e+01	2.44e+01	1.65e+04	0.34	676.23
	4	744	4.18e-02	2.18e+01	2.18e+01	1.45e+04	0.46	665.14
	23	8064	1.29e-03	3.07e+00	3.08e+00	2.39e+03	3.81	775.97
	24	8448	1.09e-03	2.81e+00	2.81e+00	2.19e+03	3.94	779.36
	25	8832	9.09e-04	2.56e+00	2.57e+00	2.00e+03	4.02	778.21

Table 5.11: Example 4: DG methods

k	level	NDOF	$\|u - \tilde{u}_h\|_0$	$\|\mathbf{q} - \mathbf{q}_h\|_0$	$\|(\mathbf{q},u)-(\mathbf{q}_h,u_h)\|$	η	EOC	EI
				H-NC methods				
1	1	8	5.16e-01	4.34e+01	4.35e+01	4.79e+02	0.00	11.01
	2	25	3.28e-01	2.75e+01	2.75e+01	4.21e+02	0.80	15.31
	3	81	3.63e-01	2.80e+01	2.81e+01	5.63e+02	-0.04	20.04
	4	149	2.25e-01	2.75e+01	2.76e+01	5.79e+02	0.06	20.98
	27	24088	4.52e-03	7.98e+00	8.00e+00	1.50e+02	0.83	18.75
	28	28299	3.92e-03	7.29e+00	7.31e+00	1.38e+02	1.12	18.88
	29	33657	3.67e-03	6.81e+00	6.83e+00	1.29e+02	0.78	18.89
5	1	40	1.60e-01	1.83e+01	1.85e+01	4.00e+03	0.00	216.22
	2	180	8.19e-02	1.96e+01	1.97e+01	2.84e+03	-0.08	144.16
	3	455	5.24e-02	1.84e+01	1.86e+01	2.59e+03	0.12	139.25
	4	760	3.91e-02	1.72e+01	1.73e+01	2.43e+03	0.28	140.46
	33	9460	2.53e-04	1.24e+00	1.25e+00	2.19e+02	5.69	175.20
	34	9760	2.12e-04	1.13e+00	1.15e+00	2.01e+02	5.34	174.78
	35	10060	1.78e-04	1.04e+00	1.05e+00	1.85e+02	6.01	176.19
7	1	56	1.14e-01	1.34e+01	1.37e+01	5.31e+03	0.00	387.59
	2	252	6.24e-02	1.52e+01	1.54e+01	3.76e+03	-0.16	244.16
	3	637	4.14e-02	1.44e+01	1.46e+01	3.41e+03	0.12	233.56
	4	1064	3.16e-02	1.34e+01	1.35e+01	3.18e+03	0.31	235.56
	33	13244	2.02e-04	9.36e-01	9.54e-01	2.77e+02	5.36	290.36
	34	13664	1.70e-04	8.56e-01	8.73e-01	2.53e+02	5.68	289.81
	35	14084	1.42e-04	7.83e-01	7.99e-01	2.32e+02	5.85	290.36

Table 5.12: Example 4: NC method

Chapter 6

Conclusion

In this thesis, we analyzed hybrid mixed methods, hybridized discontinuous Galerkin methods, as well as hybridized H^1-conforming and nonconforming finite element methods for the numerical approximation of the solution of the linear second order reaction diffusion equation. We introduced a residual based a posteriori error estimator, which consists of an estimation of the consistency error of the two components of the approximate solution in their respective spaces, and element residuals, with respect to the first and the second equation of the primal mixed variational formulation. Depending on the regularity of the discrete spaces of the actual hybrid method, one or the other estimator term may vanish a priori. We proved the reliability of the error estimator, which is valid for all individual hybrid methods under consideration. For the hybridized interior penalty discontinuous Galerkin methods and the local discontinuous Galerkin methods, it could also be proved that the estimator can be bounded from above by the actual error and data oscillation. Thus, for a fixed polynomial degree, the estimator is not only reliable but also efficient. Finally, we documented a series of numerical tests, confirming the theoretical results.

For the hybridized local discontinuous Galerkin method, using a suitably chosen penalty parameter, there is numerical evidence that, actually, the efficiency bound may be independent of the polynomial degree. This property, together with the algorithmic flexibility of the hybrid methods, makes the hybridized local discontinuous Galerkin method a possible candidate for an efficient h-p adaptive finite element method. However, the efficiency proof of an a posteriori error estimator up to constants, that do not depend on the polynomial degree of the discrete function spaces, seems to be a challenging task.

Dank

Schließlich möchte ich es nicht vermissen, einigen Personen meinen Dank auszusprechen, die zum Gelingen dieser Arbeit beigetragen haben.

An erster Stelle danke ich Prof. Dr. Ronald H. W. Hoppe für die stets aktuellen und zielführenden Anregungen und für die kontinuierliche und vertrauensvolle Unterstützung, die mir das Erstellen dieser Arbeit erst ermöglicht hat.

Ich danke den Kollegen am Lehrstuhl für Angewandte Analysis, insbesondere Prof. Malte Peter, Christian Möller, Isabell Graf, Alexandra Gaevskaya, Oleg Boyarkin, Christopher Linsenmann und Thomas Fraunholz, dass sie sich stets Zeit für Diskussionen zu mathematischen Fragen im Umfeld meiner Arbeit genommen haben.

Darüberhinaus gilt mein Dank meinen Eltern und Geschwistern, die mich insbesondere in den letzten Monaten immer unterstützt und bekräftigt haben.

Bibliography

[1] M. Ainsworth. A synthesis of a posteriori error estimation techniques for conforming, non-conforming and discontinuous Galerkin finite element methods. in: Z.-C. Shi (ed.) et al., Recent advances in adaptive computation. Providence, RI: AMS, 1-14, 2005.

[2] M. Ainsworth and J. Oden. *A posteriori error estimation in finite element analysis.* Pure and Applied Mathematics. A Wiley-Interscience Series of Texts, Monographs, and Tracts. Chichester: Wiley , 2000.

[3] A. Alonso. Error estimators for a mixed method. *Numer. Math.*, 74(4):385–395, 1996.

[4] D. Arnold and F. Brezzi. Mixed and nonconforming finite element methods: Implementation, postprocessing and error estimates. 1985.

[5] D. N. Arnold. An interior penalty finite element method with discontinuous elements. *SIAM J. Numer. Anal.*, 19:742–760, 1982.

[6] D. N. Arnold, F. Brezzi, B. Cockburn, and L. Marini. Unified analysis of discontinuous Galerkin methods for elliptic problems. *SIAM J. Numer. Anal.*, 39(5):1749–1779, 2002.

[7] I. Babuška. Error-bounds for finite element method. *Numerische Mathematik*, 16:322–333, 1971. 10.1007/BF02165003.

[8] W. Bangerth and R. Rannacher. *Adaptive finite element methods for differential equations.* Lectures in Mathematics, ETH Zürich. Basel: Birkhäuser. , 2003.

[9] A. Bonito and R. H. Nochetto. Quasi-optimal convergence rate of an adaptive discontinuous Galerkin method. *SIAM J. Numer. Anal.*, 48(2):734–771, 2010.

[10] D. Braess and R. Verfürth. A posteriori error estimators for the Raviart-Thomas element. *SIAM J. Numer. Anal.*, 33(6):2431–2444, 1996.

[11] J. H. Bramble and J. Xu. A local post-processing technique for improving the accuracy in mixed finite-element approximations. *SIAM J. Numer. Anal.*, 26(6):1267–1275, 1989.

[12] F. Brezzi, J. j. Douglas, and L. Marini. Two families of mixed finite elements for second order elliptic problems. *Numer. Math.*, 47:217–235, 1985.

[13] F. Brezzi and M. Fortin. *Mixed and hybrid finite element methods.* Springer Series in Computational Mathematics. New York etc.: Springer-Verlag. , 1991.

[14] C. Carstensen. A posteriori error estimate for the mixed finite element method. *Math. Comput.*, 66(218):465–476, 1997.

[15] C. Carstensen. A unifying theory of a posteriori finite element error control. *Numer. Math.*, 100(4):617–637, 2005.

[16] C. Carstensen and S. Bartels. Each averaging technique yields reliable a posteriori error control in FEM on unstructured grids. I: Low order conforming, nonconforming, and mixed FEM. *Math. Comput.*, 71(239):945–969, 2002.

[17] C. Carstensen, M. Eigel, C. Löbhard, and R. Hoppe. A Review of Unified A Posteriori Finite Element Error Control. 1(1):1–1, 2010.

[18] C. Carstensen, T. Gudi, and M. Jensen. A unifying theory of a posteriori error control for discontinuous Galerkin FEM. *Numer. Math.*, 112(3):363–379, 2009.

[19] C. Carstensen, J. Hu, and A. Orlando. Framework for the a posteriori error analysis of nonconforming finite elements. *SIAM J. Numer. Anal.*, 45(1):68–82, 2007.

[20] J. Cascon, C. Kreuzer, R. H. Nochetto, and K. G. Siebert. Quasi-optimal convergence rate for an adaptive finite element method. *SIAM J. Numer. Anal.*, 46(5):2524–2550, 2008.

[21] P. Castillo, B. Cockburn, I. Perugia, and D. Schötzau. An a priori error analysis of the local discontinuous Galerkin method for elliptic problems. *SIAM J. Numer. Anal.*, 38(5):1676–1706, 2000.

[22] B. Cockburn. Discontinuous Galerkin methods. *ZAMM, Z. Angew. Math. Mech.*, 83(11):731–754, 2003.

[23] B. Cockburn and J. Gopalakrishnan. A characterization of hybridized mixed methods for second order elliptic problems. *SIAM J. Numer. Anal.*, 42(1):283–301, 2004.

[24] B. Cockburn and J. Gopalakrishnan. Error analysis of variable degree mixed methods for elliptic problems via hybridization. *Math. Comput.*, 74(252):1653–1677, 2005.

[25] B. Cockburn, J. Gopalakrishnan, and R. Lazarov. Unified hybridization of discontinuous Galerkin, mixed, and continuous Galerkin methods for second order elliptic problems. *SIAM J. Numer. Anal.*, 47(2):1319–1365, 2009.

[26] B. Cockburn, J. Gopalakrishnan, and F.-J. Sayas. A projection-based error analysis of HDG methods. *Math. Comput.*, 79(271):1351–1367, 2010.

[27] M. Crouzeix and P.-A. Raviart. Conforming and nonconforming finite element methods for solving the stationary Stokes equations I. . *Revue francaise d' automatique, informatique, recherche operationnelle. Mathematique*, 7:33–75, 1973.

[28] B. F. de Veubeke. Displacement and equilibrium models in the finite element method. *Int. J. Numer. Methods Eng.*, 52(3):287–342, 2001.

[29] V. A. Dobrev, R. D. Lazarov, P. S. Vassilevski, and L. T. Zikatanov. Two-level preconditioning of discontinuous Galerkin approximations of second-order elliptic equations. *Numer. Linear Algebra Appl.*, 13(9):753–770, 2006.

[30] L. Donatella Marini. An inexpensive method for the evaluation of the solution of the lowest order Raviart-Thomas mixed method. *SIAM J. Numer. Anal.*, 22:493–496, 1985.

[31] W. Dörfler. A convergent adaptive algorithm for Poisson's equation. *SIAM J. Numer. Anal.*, 33(3):1106–1124, 1996.

[32] R. E. Ewing, J. Wang, and Y. Yang. A stabilized discontinuous finite element method for elliptic problems. *Numer. Linear Algebra Appl.*, 10(1-2):83–104, 2003.

[33] J. S. Hesthaven and T. Warburton. *Nodal discontinuous Galerkin methods. Algorithms, analysis, and applications.* Texts in Applied Mathematics 54. New York, NY: Springer. xiv, 500 p. , 2008.

[34] R. Hoppe, G. Kanschat, and T. Warburton. Convergence analysis of an adaptive interior penalty discontinuous Galerkin method. *SIAM J. Numer. Anal.*, 47(1):534–550, 2009.

[35] R. H. Hoppe and B. Wohlmuth. Multilevel iterative solution and adaptive mesh refinement for mixed finite element discretizations. *Appl. Numer. Math.*, 23(1):97–117, 1997.

[36] P. Houston, D. Schötzau, and T. P. Wihler. Energy norm a posteriori error estimation of hp-adaptive discontinuous Galerkin methods for elliptic problems. *Math. Models Methods Appl. Sci.*, 17(1):33–62, 2007.

[37] G. Kanschat and R. Rannacher. Local error analysis of the interior penalty discontinuous Galerkin method for second order elliptic problems. *J. Numer. Math.*, 10(4):249–274, 2002.

[38] O. A. Karakashian and F. Pascal. Convergence of adaptive discontinuous Galerkin approximations of second-order elliptic problems. *SIAM J. Numer. Anal.*, 45(2):641–665, 2007.

[39] K. Y. Kim. A posteriori error analysis for locally conservative mixed methods. *Math. Comput.*, 76(257):43–66, 2007.

[40] I. Kossaczký. A recursive approach to local mesh refinement in two and three dimensions. *J. Comput. Appl. Math.*, 55(3):275–288, 1994.

[41] C. Lovadina and R. Stenberg. Energy norm a posteriori error estimates for mixed finite element methods. *Math. Comput.*, 75(256):1659–1674, 2006.

[42] J. Melenk and B. Wohlmuth. On residual-based a posteriori error estimation in hp-FEM. *Adv. Comput. Math.*, 15(1-4):311–331, 2001.

[43] P. Morin, K. G. Siebert, and A. Veeser. A basic convergence result for conforming adaptive finite elements. *Math. Models Methods Appl. Sci.*, 18(5):707–737, 2008.

[44] P. Neittaanmäki and S. Repin. *Reliable methods for computer simulation. Error control and a posteriori estimates.* Studies in Mathematics and its Applications. Amsterdam: Elsevier , 2004.

[45] S. Nicaise and E. Creusé. Isotropic and anisotropic a posteriori error estimation of the mixed finite element method for second order operators in divergence form. *ETNA, Electron. Trans. Numer. Anal.*, 23:38–62, 2006.

[46] R. Owens. Spectral approximations on the triangle. *Proc. R. Soc. Lond., Ser. A, Math. Phys. Eng. Sci.*, 454(1971):857–872, 1998.

[47] L. Payne and H. Weinberger. An optimal Poincaré inequality for convex domains. *Arch. Ration. Mech. Anal.*, 5:286–292, 1960.

[48] A. Quarteroni and A. Valli. *Numerical approximation of partial differential equations.* Springer Series in Computational Mathematics. Berlin: Springer-Verlag, 1994.

[49] P. Raviart and J. Thomas. A mixed finite element method for 2nd order elliptic problems. A. Dold (ed.) et al., Math. Aspects Finite Elem. Meth., Proc. Conf. Rome 1975, Lect. Notes Math. 606, 292-315, 1977.

[50] S. Repin, S. Sauter, and A. Smolianski. Two-sided a posteriori error estimates for mixed formulations of elliptic problems. *SIAM J. Numer. Anal.*, 45(3):928–945, 2007.

[51] B. Rivière. *Discontinuous Galerkin methods for solving elliptic and parabolic equations. Theory and implementation.* Frontiers in Applied Mathematics 35. Philadelphia, PA: Society for Industrial and Applied Mathematics (SIAM) , 2008.

[52] J. Roberts and J.-M. Thomas. Mixed and hybrid methods. Ciarlet, P.G. (ed.) et al., Handbook of numerical analysis Vol. II: Finite element methedos (Part 1). Amsterdam: Elsevier, 1991.

[53] R. Stenberg. Postprocessing schemes for some mixed finite elements. *RAIRO, Modélisation Math. Anal. Numér.*, 25(1):151–167, 1991.

[54] L. Tartar. *An Introduction to Sobolev Spaces and Interpolation Spaces.* Lecture Notes of the Unione Matematica Italiana 3. Berlin: Springer., 2007.

[55] R. Verfürth. *A review of a posteriori error estimation and adaptive mesh-refinement techniques.* Wiley-Teubner Series Advances in Numerical Mathematics. Chichester: John Wiley & Sons. Stuttgart: B. G. Teubner. vi, 1996.

[56] R. Verfürth. On the constants in some inverse inequalities for finite element functions. Technical report, Ruhr-Universität Bochum, 2004.

[57] M. Vohralík. A posteriori error estimates for lowest-order mixed finite element discretizations of convection-diffusion-reaction equations. *SIAM J. Numer. Anal.*, 45(4):1570–1599, 2007.

[58] M. Vohralík. Unified primal formulation-based a priori and a posteriori error analysis of mixed finite element methods. *Math. Comput.*, 79(272):2001–2032, 2010.

[59] T. Warburton and J. S. Hesthaven. On the constants in hp-finite element trace inverse inequalities. *Comput. Methods Appl. Mech. Eng.*, 192(25):2765–2773, 2003.

[60] B. I. Wohlmuth and R. H. Hoppe. A comparison of a posteriori error estimators for mixed finite element discretizations by Raviart-Thomas elements. *Math. Comput.*, 68(228):1347–1378, 1999.

Lebenslauf

Name   Johannes Neher

Ausbildung

2001	Abitur am Maristenkolleg Mindelheim
2002 – 2007	Studium Lehramt an Gymnasien für die Fächer Mathematik und Physik, Universität Augsburg
2008 – 2011	Promotion am Lehrstuhl für Angewandte Analysis mit Schwerpunkt Numerische Mathematik der Universität Augsburg